ENCYCLOPEDIA OF THE

WEATHER

ENCYCLOPEDIA OF THE

WEATHER

IAN WESTWELL

CONTENTS

INTRODUCTION

Humankind has always had an uneasy relationship with the weather and climate. In the earliest times, weather and climate and the forces that create them, were not understood, being ascribed to the work of some unseen, all-powerful divinity. Early man lived in fear of his environment, not able to comprehend its true nature. All of humankind's activities, especially agriculture, were conducted in sympathy with seasonal changes in the weather, which influenced the timing and length of the growing season. It is no coincidence that one of the major catastrophes encountered in the Bible is the great flood described in the book of Genesis.

However, as humankind began to enquire into the natural world, both weather and climate, and the processes that influence them, were studied and our understanding of them grew. Our ancestors no longer lived in such fear of the weather and were able to develop strategies to make the best of its advantageous features and minimize its more potentially dangerous aspects.

In the modern world, we do not fully comprehend certain aspects of the weather and climate, but our understanding of the process involved is greater than it has ever been before. Man has attempted to control and modify the weather directly and deliberately, through such techniques as cloud seeding to produce rain, and has also influenced climate both indirectly and not always beneficially. Industrialization since the mid-1800s, particularly the burning of fossil fuels, the production of aerosols and chlorofluorocarbons, and the destruction of the tropical rainforests, has had a possibly long-term and detrimental influence on the Earth's climate, chiefly through the greenhouse effect and global warming. No one knows for certain if global warming is irreversible or precisely what impact it will have on the planet, but many commentators suggest that low-lying areas will be flooded as polar ice melts, raising sea levels, and that certain illnesses, especially some types of cancer,

LEFT: *Aristotle — his observations on the weather in* Meteorologica *would be unequaled for nearly 2,000 years and the book title would give us the word "meteorology."*

RIGHT: *Man's influence on the weather has increased exponentially since the Industrial Revolution. The burning of fossil fuels, the production of aerosols and the destruction of rainforests has had a long term effect on our climate.*

PREVIOUS PAGE: *Lightning — to early man the product of the gods; today, an occurrence caused when positive and negative charges within a cloud become so strong that they can overcome the natural resistance of the air and discharge (see page 111).*

ABOVE: *Evangelista Torricelli, the seventeenth century Italian physicist who experimented with vacuums and pressure. In 1643 he proved the existence of air pressure and gave the first description of the barometer (a "torricellian" tube) to measure it.*

RIGHT: *A replica of Torricelli's "torricellian" tube, the first barometer.*

will increase. Pollution also reduces air quality and promotes increases in respiratory diseases.

It seems likely that the Chinese were the first to attempt to accurately observe the weather, some 1,100 years before the birth of Christ, but it was the ancient Greeks who made the first rigorous and scientific attempts to explain various weather features. The philosopher Aristotle (384-322BC) wrote revealing, if not altogether accurate, passages concerning features such as clouds, rain, and storms in his *Meteorologica*, which were unequaled for nearly 2000 years.

Perhaps surprisingly, the work of Aristotle and others to create a framework by which the weather and climate could be understood was not progressed for a considerable period. In both the Dark and Middle Ages, the western world was beset by a number of disasters, mainly involving war, plague, famine, and disease. Not being sufficiently scientifically advanced to identify rational causes or explanations for these events, those beset by them turned to the supernatural. Unfortunately, the weather, and the occasionally damaging events associated with it, were also considered as supernatural and the work of the devil. Those who might wish to study the weather were in danger of being identified as disciples of Satan.

The transition toward a more rational study of the weather probably began in the early sixteenth century, partly based on the simple, often witty, sayings, such as "A mackerel sky and a heavily made-up lover; one's as short lived as the other," which reflect the experience of those, including farmers and sailors, who had amply reason to be attuned to changes in the weather. However, the development of the science of meteorology really began from the middle of the sixteenth century. For more than 100 years, numerous scientific advances that we now consider commonplace were made. In 1643, for example, Evangelista Torricelli, showed the existence of air pressure and created the barometer to measure it accurately.

Great steps were also being made in the recording of temperature during this period. The first sealed thermometer, based on work undertaken by the astronomer Gallileo

ABOVE RIGHT: *Anders Celsius, the Swedish astronomer who devised the Centigrade or Celsius scale, where 0° is the freezing and 100° the boiling points of water.*

BELOW RIGHT: *Antoine Lavoisier, who developed the first guidelines for predicting the weather.*

Gallilei at the end of the sixteenth century, was created for Grand Duke Ferdinand of Tuscany in 1641 and a recognized series of temperature grades based on the boiling and freezing points of water were suggest by Italian physicist Carlo Renaldini in 1694. In 1714, the German physicist Daniel Fahrenheit invented the temperature scale that continues to bear his name and used mercury to record temperature change for the first time; less than 30 years later, Swedish astronomer Anders Celsius devised another temperature scale.

Progress was also being made in creating a methodical system for recording the weather during this period. In the middle of the eighteenth century, Frenchman Antoine Lavoisier developed the first guidelines for predicting the weather, including the measurement of air pressure by barometer, the recording of wind speed and direction, and atmospheric mois-

ture. The first practical hygrometer, a device to record humidity, was build by the Swiss physicist Horace de Saussure in about 1780.

Impetus for the creation of a national meteorological service grew out of a series of damaging weather episodes that afflicted France in the first half of the nineteenth century. In 1821, a leading member of the French government issued a statement requesting local officials throughout the country to study what he termed "aberrations." Many suggestions, some sensible, other barely credible, to explain the "aberrations" were put forward by the officials, but the most sensible proposal suggested that only the daily recording of the weather would enable the causes of the "aberrations" to be identified.

The French government redoubled its attempts to understand and possibly predict the weather following a disastrous event that occurred during the Crimean War. In 1854, the French Navy lost close to 40 vessels in a violent storm and the director of the Paris observatory, Urbain Le Verrier, was asked to investigate the matter. He reported that the onset of the destructive storm could have been predicted and that weather forecasting had to be modernized. His suggestions were accepted by the government, which established a network of 24 recording stations, many of which were connected by telegraph. Gradually, it became apparent that the key to successful weather prediction was the study in changes in air pressure.

A key figure in developing our understanding of the relationship between changes in air pressure and wind direction was a Dutch scientist, Christoph Buys Ballot, who correctly stated in 1857 that winds always flow along isobars from areas of high pressure to zones of low pressure. By measuring barometric pressure at various points (weather stations), it was then possible to connect places with equal barometric pressure by drawing isobars on a

map. Each station would also give readings of other phenomena, such as temperature, cloud cover, precipitation, and so forth. It was then possible to give an overview of the weather experienced over a wide area, and knowing the probably direction and speed of the winds, suggest the weather that should be expected at a particular place at a later time.

Other nations followed the French lead and established their own meteorological service during the century. The United States, for example, created a weather service under the control of the Army Signal Corps in 1870. However, these national networks were essentially isolated from each other and what was urgently required was an international system of weather recording and forecasting. The meeting of the Second International Meteorological Congress suggested the creation of a nine-man committee to coordinate and standardize the recording of the weather and the exchange of information between the relevant international bodies.

Although the advances made in gathering meteorological data made by individuals such as Buys Ballot were undoubtedly of great significance, the gathering systems had two serious drawbacks: the recordings were always made at ground level and did not take any account of the impact of terrain on the weather. It became vital to analysis conditions in the upper reaches of the lower atmosphere and a number of new instruments were developed to carry out the task. Frenchman Leon Teisserenc de Bort was one of the first meteorologists to carrying out soundings of the lower atmosphere by using kites and balloons to carry recording devices aloft. There were problems with these devices; information could only be gathered when the kites or balloons were brought back to earth and many were never recovered, having been blown way off course by the wind. However, this problem was overcome at the end of the 1920s, when two scientists, Frenchman Robert Bureau and Russian Pavel Malchanov, created the radiosonde. A balloon with recording devices, it also carried a transmitter to radio data to a recording station.

By the close of the first half of the twentieth century, most of the devices used today to forecast the weather had been developed, but modern forecasting has become increasingly sophisticated, particularly due to the advent of satellites able to record weather patterns from space. These have not replaced the older technologies, but have added a new level of accuracy to the forecast that we hear, see, and read every day.

FORECASTING

The essential element that marks out modern meteorology from historical attempts to record weather and its variations over time is the use of data to predict future conditions with a supposedly high degree of accuracy. Modern weather forecasting systems are undoubtedly more accurate, faster, and have a greater technological sophistication than any that have gone before, thereby allowing meteorologists to estimate future conditions with a degree of confidence. High-speed computers and geostationary or orbiting satellites are now used routinely to find the position and track the movement of major weather systems, and then predict their future paths.

However, it needs to be borne in mind that modern forecasters are not so dissimilar from those who practised the science in the past. Forecasting, both old and new, has always consisted of three key interrelated elements — data gathering and analysis, prediction, and dissemination of information — and the accuracy of forecasts still relies to some extent on the skills and experience of trained meteorologists, only the tools of the trade have changed down the decades. Experienced meteorologists must also be able to step back from the large pattern of weather and identify more small-scale variations from the overall situation brought about by local factors, such as terrain or the relationship between land and water. Computer-generated models of the weather based on — hopefully — accurate recordings only indicate how the weather should develop on the basis of the weather-pattern models that run on the computer and do not necessarily bear any relation to the "real" events that unfold.

The development of sophisticated predictive work in modern meteorology really begins with the theoretical work of a Norwegian scientist, Vilhelm Bjerknes, in the early years of the twentieth century. Bjerknes postulated that by applying the laws of thermodynamics and fluid mechanics to the Earth's atmosphere it should be possible to predict the nature of the atmosphere at some future date. The Norwegian thus believed that the laws of physics could be applied successfully to the study of the dynamics of the atmosphere and that the evolution of weather systems could be modeled with a sufficiently high degree of accuracy. Perhaps unfortunately for Bjerknes, however, the computations involved in such modeling were so complex that the technology needed to carry out both the recording of data and the identification of future weather patterns was not available at the time he proposed his new approach to meteorology. Nevertheless, the Norwegian had laid the foundations for a rational, scientific approach to the study of meteorology.

With regard to gathering data on the prevailing weather, the modern world is well served by recording stations, some more sophisticated than ever before; others which would not be especially unfamiliar to a meteorologist from an earlier time. Meteorologists are now able to gather information from the ground, the atmosphere, and space. The globe is covered by a network of some 10,000 permanent ground stations and some 5,000 weather ships, which record daily changes in air pressure, precipitation, wind, and temperature. These manned stations at ground level are backed up by a host of automatic weather recording devices, chiefly unmanned stations and weather buoys, from where information can be transmitted or beamed back to base for immediate analysis.

Science has also devised many new recording technologies. Over 1,000 radiosonde balloons, a device first developed in the late 1920s, are released into the atmosphere every day and transmit details of upper atmosphere conditions back to ground stations. Many of the world's leading nations also fund specially converted aircraft which are able to monitor weather conditions over large distances. However, the greatest leap forward in weather data gathering in the fairly recent past has been the use of satellites.

Satellites have two great advantages over other forms of meteorological data-gathering systems in that they provide a continuous, rather than snapshot, picture of the atmosphere's condition and can provide information on areas of much greater geographical extent than other recording stations, thereby identifying large-scale weather systems, such as thunderstorms and hurricanes, whose path they can also follow.

There are two types of satellite. Geostationary types hold a fixed position above the Earth, usually at a height of around 22,000 miles above the surface at the equator, and each is able to provide images of approximately one-third of the surface. Polar-orbiting satellites, found at heights between 500 and

PREVIOUS PAGE: *What will it be like tomorrow? The essential element that marks out modern meteorology from historical attempts to record weather and its variations over time is the use of data to predict future conditions with a supposedly high degree of accuracy.*

BELOW: *Weather satellites allow observation not only of local weather systems, but remote areas where surface meteorological data is lacking. Often in polar orbit, they measure temperatures — surface and atmospheric — cloud cover, water vapor and atmospheric energy fluxes as well as being particularly useful in major storm direction assessment.*

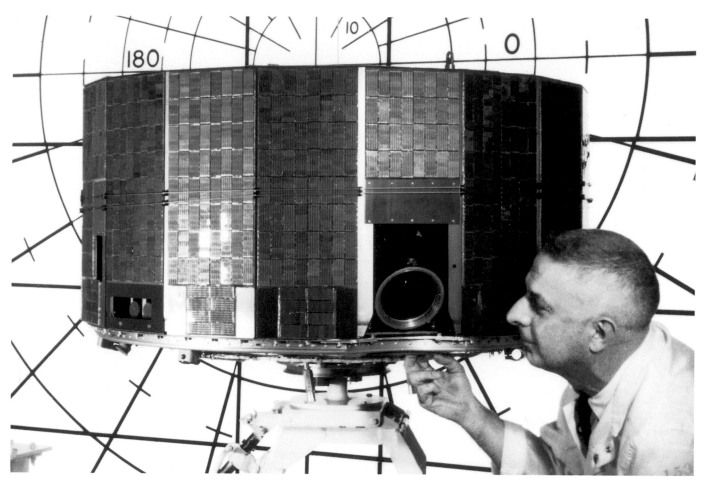

ABOVE: *The TIROS (Television Infra-red Observational Satellite) series of meteorological satellites started in 1960 as an experimental sequence of ten "metsats." This pioneering system sent back thousands of cloud cover pictures and was the forerunner of the increasingly sophisticated systems in use today. TIROS 1 had a 78 day life during which it took over 20,000 photographs. This photograph is dated 1961.*

LEFT: *By 1978 TIROS had reached the fourth generation. Here TIROS N — NASA's prototype for the new generation of "metsats" — is undergoing ground tests at NASA. It would launch on October 13, 1978. The first operational satellite in the series — the United States' NOAA (National Oceanic and Atmospheric Administration) 6 satellite — was launched eight months later.*

around 600 miles above the surface, circle the globe in a fixed orbit but because the Earth itself is rotating, the images they transmit move progressively westward through the day. Both types have advantages and disadvantages for the meteorologist. Geostationary types, because of their altitude, "see" greater areas of the surface but the images they transmit are not as sharp as those provided by the lower-altitude polar-orbiting satellites. However, while the polar types provide crisper images, their coverage of the surface at any one time is of lesser geographical extent.

Satellites gather information on the atmosphere by using two types of sensors which look at visible, reflected solar radiation or infrared radiation. Sensors studying visible radiation provide black-and-white photographic images on which features such as land, water, and cloud, can be identified because of their different reflective qualities. Infrared sensors measure the temperature of objects so, for example, low clouds, which are "hotter," can be distinguished from high ones, which are cold.

The gathering and processing of meteorological information is coordinated at national and international level. The first attempt to link various national meteorological departments together occurred in 1878, when the International Meteorological Organization was established. This evolved into the World Meteorological Organization (W.M.O.), part of the United Nations, in 1951. Today, the W.M.O. is based in Geneva, Switzerland, and processes information provided by more than 175 countries around the world.

Data gathered by the various individual recording stations of these countries at set times of the day are transmitted to a trio of World Meteorological Centers situated in Melbourne, Moscow, and Washington DC. At each of these, the data is synthesized to produce maps indicating the state of the atmosphere at the time the local recordings were made. By studying these maps, meteorologists attempt to predict future weather conditions, drawing up predictive charts that are returned to more than 30 Regional Specialized Meteorological Centers, which then send the information to the National Centers for Environmental Prediction in each of the W.M.O.'s member states. All data collected and distributed by the W.M.O. passes through its Global Telecommunications system. Consequently, weather forecasters are able to produce reasonably accurate weather maps at high speed and disseminate the information to other meteorologists and the public at high speed. This is, of course, vital as reporting yesterday's weather is about as useful as reading yesterday's newspapers to most people.

The various weather agencies, once they have gathered all the available weather data produce two types of weather map. Surface weather maps are drawn at intervals of three to six hours in the northern hemisphere and the readings shown are always adjusted to sea level, while upper-atmosphere maps display information gathered by radiosonde balloons. It is on the basis of the information displayed on these two types of chart that meteorologists make their weather forecasts.

As has already been noted, synthesizing all of the data provided by the various recording devices is incredibly complex. Early attempts, such as that made during World War I by the English meteorologist Lewis Fry Richardson, to calculate future weather failed because the technology of the time was not equal to the task. Nevertheless, Richardson inspired others to continue this mathematical approach to forecasting. Swedish meteorologist Carl-Gustav Rossby was able to simplify some of Richardson's calculations in 1939, but it was the development of computer technology that finally opened the door to numerical forecasting.

In 1950, an American mathematician, John von Neumann employed at Princeton's Institute for Advanced Study, published details of his work in numerical forecasting by computer. Using the massive, 30-ton ENIAC digital computer, he had been able to create models of the weather based on the laws of thermodynamics and fluid mechanics as had been suggested by the Norwegian meteorologist Vilhelm Bjerknes earlier in the century. Within the space of a few years, the use of computers as weather predictors became widespread and, as computers became increasingly sophisticated, the quality of the information provided improved almost immeasurably. However, to make these theoretical model of true value to meteorologists and forecasters, they had to be programmed with actual data.

To achieve a degree of accuracy the Earth is now divided into a box-like network, each of which is further divided into a number of ver-

RIGHT: *The anemometer measures windspeed and is one of the key instruments for the weather forecaster. There are three types — the usual windmill (as here) using cups or shaped vanes; the rate of cooling of an electrically heated wire; an open-ended tube facing the airflow (the pitot tube used on aircraft).*

tical divisions. Computers then process the ground-level and atmospheric data provided by the various recording stations and ascribe weather details (pressure, temperature, humidity, for example) to each section of the box-like grid, thereby creating a snapshot of the weather at the time the initial readings were made. As the weather is considered to behave on certain set scientific principles, it is then possible to suggest accurately the ways in which it will develop in the near future — at least in theory. So, for example, information on the development of a storm can be used to suggest the way in which it will evolve and in what direction it will move. Clearly such information is of inordinate value in the case of highly destructive weather features, such as hurricanes.

While there is no doubting that modern weather forecasting is markedly superior to anything that has gone before, it is far from infallible. The quality of a forecast still depends on the accuracy of the data provided by various gathering stations and, while recording devices are increasingly precise, they are not perfect. Also, the data-gathering network, while extensive, is not universal. The northern hemisphere is covered extensively by weather stations, but there are noteworthy gaps in the southern hemisphere. Consequently, forecasts in certain parts of the globe are based on "guess-estimates" which may bear only a passing relationship to the actual conditions.

Modern meteorologists feel fairly confident that they can predict weather conditions up to a week or so ahead with a fairly high degree of accuracy. The United States' National Weather Service, probably the most sophisticated meteorological agency in the world, estimates that over 80 percent of its one-day forecasts are accurate. Much of this improvement in the quality of forecasts is due to the employment of more sophisticated computers,

the greater quality of the information provided by weather stations, the development of a coordinated global meteorological network, and advances in our understanding of the mechanics of the atmosphere. However, our understanding of the mechanics of the atmosphere and our ability to interpret data is still insufficient to predict the weather with any degree of accuracy beyond seven days. Over longer periods the quality of a forecast deteriorates significantly and beyond 10 days rarely matches the weather actually experienced.

However, some agencies attempt to generate long-term weather forecasts, usually called outlooks, for conditions 30 or even 90 days ahead. Generally, the outlooks are much less detailed than shorter-term forecasts and tend to concentrate on identifying patterns of temperature or precipitation that may vary significantly from the expected norm. To give some indication of accuracy, the Long-Range Prediction Branch of the National Weather Service's Climate Analysis Center in the United States has estimated that on a scale of zero to 100, in which zero represents the likelihood of estimating the actual temperature or precipitation by chance and 100 equals a perfect forecast, then the accuracy of temperature forecasts would score eight and precipitation a mere four. Clearly, the quality of long-range forecasting is far from high.

Indeed, some commentators have suggested that the belief that the atmosphere behaves in certain predictable ways may have been exaggerated and that, if this hypothesis is true, then the whole basis of modern numerical forecasting may be fundamentally flawed. This view may well be unduly pessimistic, but may lead to new avenues of enquiry that will add to our still-imperfect knowledge of the mechanics of the weather.

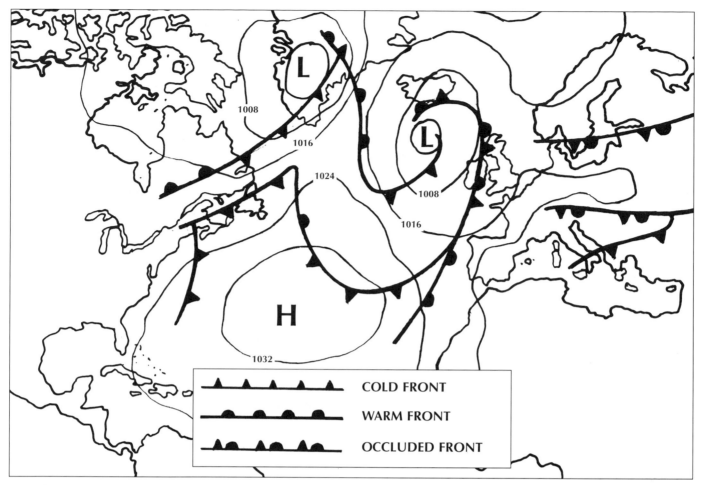

	COLD FRONT
	WARM FRONT
	OCCLUDED FRONT

Weather maps like this are published and broadcast all over the world each day. Few people do not pay an interest in the prognostications of the weather men and women. The maps are based on the surface weather maps drawn every three to six hours in the northern hemisphere and upon which forecasts are made. Also known as synoptic charts because they provide a synopsis of prevailing conditions at a particular period, such maps usually carry information relating to systems that are responsible for the weather at a particular place.

All the features shown on the map are explained in more detail in the encyclopedia. First, there is the atmospheric pressure, identified by the letters L (for Low) and H (for High). Then there are isobars — lines on the chart linking points that have equal atmospheric pressure. It is usual for the readings gathered at a weather station to be amended to sea level so that the influence of altitude on pressure is removed. On weather maps, isobars are drawn to reveal the distribution and variations in atmospheric pressure at a chosen time on a particular day in a reasonably small area, such as a country or continent. Generally, the isobars are drawn at two or four millibars on a daily weather map, and usually even numbered readings — for example 998mb, 1000mb or 1004mb and 1008mb — are shown.

Around these lows and high, the fronts move. A cold front is the dividing line between a body of advancing cold air and a mass of warm air. Clear indicators of the passage of a cold front are a drop in temperature, a veering of the wind, a rise in atmospheric pressure, and the onset of heavy showers. On certain occasions, usually when a cold front is advancing at speeds of over 25mph, a squall line develops either at the leading edge of the front or some distance ahead of it. These are associated with thunderstorms.

A warm front is the line on the Earth's surface where an advancing mass of warm air is forced to rise by the presence of a cooler, slower-moving, and denser air mass. Cloud develops as the warm air is forced aloft and precipitation is likely to take place ahead of the front. Once the front has passed, temperatures tend to rise, precipitation ends, and the wind veers.

An occluded front develops when a cold front, usually traveling at twice the speed of a warm front, catches up with the warm front.

This sort of weather map is usually accompanied by a smaller, localized forecasting map for the area (or country) identifying anticipated windspeed, temperature and weather.

18:00 29AU79 12A-Z 0006-1640 FULL DISC IR

18:00 30AU79 12A-Z 0006-1640 FULL DISC IR

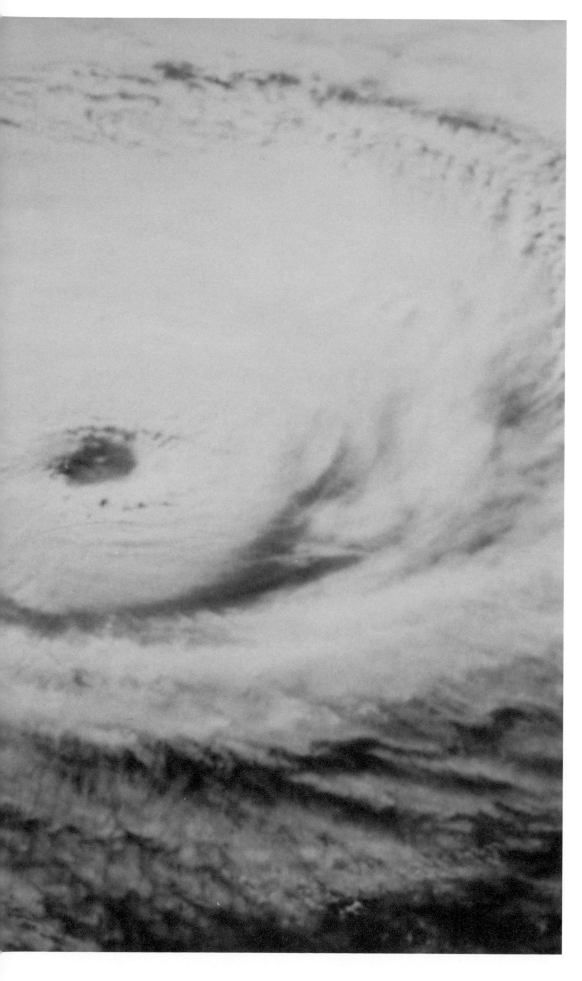

FAR LEFT: *Infrared satellite photographs of Hurricanes "David" and "Frederick" on their way toward the eastern seaboard of the United States. Predicting the track of hurricanes is always difficult, even with satellite weather information like this. Nevertheless the information can supply some advanced warning and thus save lives and livelihoods.*

LEFT: *Satellite image of Hurricane "Pat."*

THE ENCYCLOPEDIA

Absolute Drought *(Above left)*

In Great Britain, this term refers to a span of at least 15 consecutive days during which nowhere has received more than 0.2mm (0.08in) of rain. The standard is not accepted internationally.
See: Drought.

Absolute Humidity *(Below Left)*

This is the ratio, usually expressed as grams per cubic meter, between the amount of water vapor, measured in grams, to the mass, measured in kilograms, of the air containing the water vapor. This measure is not of great use in meteorology because the volume of an air mass may change, thereby leading to a change in absolute humidity, even though there is no increase or decrease in its water vapor content.
See: Air Mass, Specific Humidity, Relative Humidity, Water Vapor.

Absolute Instability

This refers to a situation in which a measure indicates that the temperature of the ambient air is falling more rapidly with altitude than 10°C per 1,000m (5.5°F per 1,000ft), the dry adiabatic lapse rate. The ambient air is considered unstable for both saturated and unsaturated air parcels under this circumstance.
See: Adiabatic Lapse Rate, Air Mass.

Absolute Stability

A layer of air, whether saturated with moisture or unsaturated, is considered stable when *any* of these three conditions are recorded: 1. temperature increases with altitude; 2. temperature does not alter with height above the ground; 3. the temperature of the ambient air falls more slowly than the moist adiabatic lapse rate as altitude increases.
See: Adiabatic Lapse Rate, Air Mass.

Accumulated Temperature

A measure often of value in measuring the effectiveness of temperature in encouraging plant growth. It is calculated over a specific period of time, usually a month, and consists of the products of time (days) and the excess of the average daily temperature above a given value. It is measured in day-degrees. In Great Britain, the basic temperature is usually 5.5°C and all temperatures above this level are conducive to plant growth in a European-type climate.
See: Growing Season, Temperature.

Acid Deposition

Acid deposition is the process by which acidic components of the air are removed from the air and left on the ground during part of the hydrological cycle. When polluted droplets reach ground level through dry deposition, they leave behind previously dissolved or suspended substances picked up when the water was rain or snow in the atmosphere. Dry deposition has two components: impaction, when aerosols are removed from the air by hitting and sticking to structures such as buildings, and gravitational settling, when particles falls to the ground under the influence of gravity. Snow and rain are usually mildly acidic as they normally dissolve carbon dioxide in the atmosphere. However, air polluted with either sulfur oxides or oxides of nitrogen reacts with moisture contained in the air to generate significant amounts of sulfuric and nitric acids. These dissolve during precipitation and their acidity is increased. If there is no precipitation, drops of sulfuric acid convert to acidic aerosols which settle on the ground through dry deposition.
See: Aerosols, Atmosphere, Gravitational Settling, Hydrological Cycle, Pollution, Precipitation.

Acid Rain

Acidity and alkalinity are measured on the pH scale, which runs from 0 to 14. When rain is tested a figure of seven indicates a neutral reading, a reading below this indicates increasing acidity, and a figure above seven indicates increasing alkalinity. Under normal circumstances, rain has a pH of 5.6, and a reading that indicates an acidity greater than this, that is, lower than 5.6, suggests acid rain. Precipitation is usually slightly acidic, but high readings are usually a rough measure of the presence of pollutants in the atmosphere.
See: Atmosphere, Pollution, Precipitation.

Adiabatic Lapse Rates

This refers to the changes in the pressure and temperature of a parcel of air, or any other gas, when heat is neither added to nor subtracted from it as it rises or falls. This means that its temperature and pressure are not affected by radiation, conduction, or mixing with its surroundings. The dry adiabatic lapse rate occurs when rising unsaturated air cools due to its expansion as pressure falls or falling air warms due to compression as pressure rises. Air

cools as it rises because atmospheric pressure decreases, decreasing the air's temperature and density as it expands. This process is known as expansional cooling. The opposite of this process is known as compressional warming. As an air parcel falls, the pressure acting on it increases and its temperature and density rise, thereby warming the air. The rate of heating or cooling that occurs as the air falls or rises is constant at a rate of 10°C per 1,000m (5.5°F per 1,000ft). The moist adiabatic lapse rate relates to the cooling of rising saturated air parcels. When rising air cools to a level at which its relative humidity reaches 100 percent and condensation or deposition occur, freeing latent heat which modifies the effect of expansional cooling, the air no longer cools at the dry adiabatic lapse rate. Rather it cools more slowly. The rate of cooling is not fixed as in the dry adiabatic lapse rate, but changes with temperature which governs the vaporization of water in the air mass. Warmer saturated air undergoing deposition or condensation gives off more latent heat to offset expansional cooling and the moist adiabatic lapse rate is smaller that that of a parcel of cool saturated air in which less latent heat is generated by deposition or condensation. The moist adiabatic lapse rate generally varies from 9°C per 1,000m (5°F per 1,000ft) for very cold saturated air to 4°C per 1,000m (2.2°F per 1,000ft) for very warm saturated air.
See: Compressional Warming, Condensation, Deposition, Expansional Cooling, Radiation, Relative Humidity.

Advection
The process by which heat is transferred between two points by the horizontal movement of air. When an air mass moves from one position to another this is known as air mass advection. Cold air advection takes place when a wind blows from a cold to a warm region, while warm air advection involves a wind blowing from a warm to a cold region.
See: Advection Fog, Air Mass, Anabatic Wind, Convection, Katabatic Wind.

Advection Fog
A type of fog which forms when a parcel of air passes horizontally over a cooler surface (ice or snow, for example) and cools to dew point.
See: Advection, Dew Point, Fog.

Aeolian
Of or relating to the wind. The word is taken from the name of the Greek god of the winds, Aeolus.
See: Wind.

Aerography
The description of the characteristics and properties of the atmosphere.
See: Atmosphere.

Aerology
The study of the atmosphere and its properties.
See: Atmosphere.

Aerosols
The atmosphere consists of not only gases but also small liquid and solid particles, known as aerosols. Water drops and ice crystals can be recognized as clouds, but other aerosols are too minute to be visible. These are usually found in the lower atmosphere, and are produced by a number of events, including human and industrial activities, particularly the burning of fossil fuels (oil and coal), volcanic eruptions, dust storms, and fires. Aerosols, although comprising a very small part of the Earth's atmosphere, are significant. Some are essential to the development of clouds, while others influence the temperature of the air by their effects on solar radiation. The increase of some aerosols due to industrialization is considered to be a key component of climatic change and global warming.
See: Atmosphere, Climatic Change, Chlorofluorocarbons, Global Warming.

Air Density
The effect of gravity on the density (mass of gas molecules per unit volume) of air. At sea level, the effect of gravity is felt most and the molecules are most closely spaced, that is denser. As the height above sea level rises, the spacing of molecules increases, air density falls, and the air thins. At a height of 10 miles, the air density is some 90 percent of that at the surface. The effects of thinning air can be seen in mountaineers who require oxygen to conquer higher peaks.
See: Atmosphere.

Air Frost
A body of air that has a temperature below 0°C (32°F).
See: Celsius, Centrigrade Scale, Temperature.

Air Mass

A very large body of air that has taken on the temperature and humidity characteristics of its region of origin. Such air masses are usually described in terms of their climatic zone of origin (polar or tropical, for example) and whether or not they originated over sea (maritime) or land (continental). So, for example, commonly occurring air masses are polar continental or tropical maritime. They can be modified as they move.
See: Air Mass Climatology, Air Mass Modification, Climatic Zones, Fronts.

Air Mass Climatology

Rather than describing the climate of an area in terms of the normal measures of climate (temperature, rainfall, and so on), their averages and extremes, air mass climatology attempts to describe the climate of an area in terms of the air masses that either develop over that area or are drawn into it by advection.
See: Advection, Air Mass, Air Mass Modification.

Air Mass Modification

These are changes that occur to an air mass as it moves from its point of origin to other areas. Changes include variations in its humidity, temperature, and stability. The causes of these are possibly threefold: the modification of an air mass as it passes over a surface and exchanges heat and moisture with it, radiational cooling, and changes generated by the vertical movement of the air.
See: Air Mass.

Air Pollution *(see illustrations on pages 32/3, 34)*

This occurs when the atmosphere contains levels of a gas or aerosol that could have potentially serious consequences for human health and the environment. Some are naturally occurring, such as sulfur dioxide, and only become a problem if their concentrations rise due to natural events. Others are generated or increased by human activity, particularly the burning of oil and coal, and the use of the internal combustion engine. In 1992, the US Environmental Protection Agency estimated that

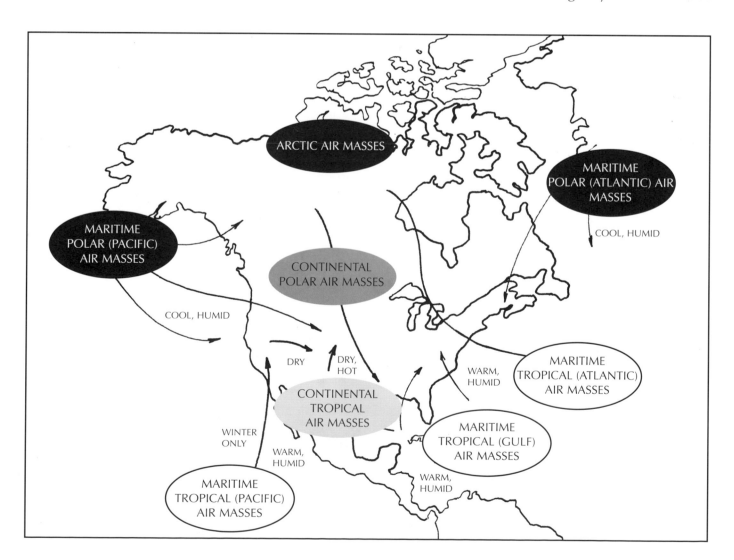

transport contributed over 83 million metric tons of pollutants to the United States, including over 63 million metric tons of carbon monoxide, a potentially lethal gas. Byproducts of such high levels of air pollution include smog.
See: Atmosphere, Carbon Monoxide, Pollution, Smog, Sulfur Dioxide.

Air Pressure

Measured in kilograms per square centimeter or pounds per square inch, this is defined as the weight per unit area of the column of air above a particular point. Pressure changes are measured by barometers and can be caused by altitude, temperature, pressure, and the interaction of air masses.
See: Air Mass, Air Pressure Tendency, Barometer.

Air Pressure Tendency

The change of air pressure at a particular place over time. A rising air pressure tendency generally indicates good weather and a falling air pressure tendency suggests stormy, worsening weather.
See: Air Pressure.

Altimeter

A barometer graduated to measure height above the ground or sea level rather than atmospheric pressure.
See: Barometer, Atmospheric Pressure.

Altocumulus *(Above Left)*

A type of cloud found at medium levels. Usually gray or white, altocumulus formations comprise an often dense mass of small and patchy clouds. They are found at heights of between 2,000 and 7,000m (6,500–23,000ft) above ground level.
See: Clouds.

Altocumulus Castellatus

Recognizable by their turreted tops, altocumulus castellatus clouds are a variety of the altocumulus type. They are a product of thundery conditions and an instability in the upper air.
See: Altocumulus, Clouds.

Altostratus *(Left)*

Occurring at medium altitudes, altostratus are flat and thick sheets of gray cloud capable of obscuring the sun under average conditions and obscuring it completely when it rains or snows.

They are usually accompanied by the passage of a depression.
See: Depression, Rain, Snow.

Anabatic Wind

A localized wind which blows up the slopes of high ground. Most common during daytime, anabatic winds are produced when a slope is heated by the sun, causing the air above it to rise by convection. Cooler air from the valley then blows up the slope to replace the rising hotter air. A katabatic wind, which occurs principally during the night, is produced by the reverse process.
See: Convection, Katabatic Wind, Wind.

Anemogram

A continuous record of variations in wind speed recorded by an anemograph. Changes in wind direction are also sometimes recorded on an anemogram.
See: Beaufort Scale.

Anemometer

A device for recording the speed and sometimes the direction of wind. Also known as a wind gauge, the most common type consists of a system of cups whose rate of rotation is recorded by a speed indicator which measures changes in wind speed over time.
See: Beaufort Scale.

Aneroidograph

A self-recording aneroid barometer.
See: Barometer.

Antarctic Ozone Hole *(Above and pages 38/9)*

An area of variable but significant size positioned over the Antarctic and found in the stratosphere which suffers serious cyclical ozone depletion. The hole grows annually between August and early October and declines in late November. While its causes and significance are a matter of on-going debate, it is thought to be caused by effects of chlorine chemical activity. Chlorofluorocarbons are the source of the chlorine.
See: Atmosphere, Chlorofluorocarbons, Global Warming, Greenhouse Effect, Ozone, Stratosphere.

Anticyclone

A region in which the atmospheric pressure is high compared with that of nearby areas. Anticyclones are usually represented on a weather map by closed and concentric isobars. Winds, frequently light nearer the anticyclone's center

and stronger on its margins, blow round an anti-cyclone in a clockwise and slightly outward direction in the northern hemisphere and anti-clockwise and slightly inward in the southern hemisphere. Anticyclones move slowly and may remain stationary for several days. Anticyclones are associated with settled weather conditions. Skies are often lacking cloud and temperatures are relatively high in summer, while in winter the lower levels are cooled and associated with fog. Most anticyclones are shortlived, but there are two permanent areas of anticyclones chiefly situated over the worlds oceans at about 30 degrees north and 30 degrees south. They vary in extent and power with the changing seasons and move a little north and south.
See: Cyclone, Depression, Front, Horse Latitudes, Isobar.

Anti-Trade Winds

These winds are found in the upper air above the trade winds and blow in the opposite direction. So for example, a northeast trade wind at the surface will be matched by a southwest anti-trade wind in the upper air. It is thought that the anti-trade winds move air transported toward the equator by lower level trade winds back into the upper atmosphere.
See: Trade Wind, Wind.

Arctic Air

Very cold and dry mass of air that chiefly develops in winter over the north interior of the North American continent, Greenland, and the Arctic basin. It is signified on charts by the letter A.
See: Air Mass.

Arctic High

An anticyclone that develops in areas dominated by cold and dry Arctic air.
See: Anticyclone, Arctic Air.

Arctic Smoke

A type of fog produced when bitterly cold air blows off the land and then condenses over warmer adjacent seas.
See: Condensation, Fog, Steam Fog.

Argon

An inert gas that makes up 0.93 percent of the Earth's atmosphere by volume.
See: Atmosphere.

Arid

Severely lacking in rain. Arid is usually applied to areas with little vegetation or climates with less than 25.4cm (10 inches) of rainfall per year.
See: Climatic Zones.

Atmosphere

This is the thin envelope surrounding the Earth which contains a mixture of gases, dust particles, pollutants, and water vapor. It is in the lower portions of the atmosphere where weather and climatic change take place. Its chief functions are to shield life from dangerous ultraviolet radiation from space, supply water without which life could not exist, and contain the gases required for breathing and photosynthesis. The present atmosphere and its components represent more than 4.6 billion years of evolution. The principal component gases in the lower atmosphere are: nitrogen (78.08 per cent by volume); oxygen (20.95 percent); argon (0.93 percent); carbon dioxide (0.035 percent); neon (0.0018 percent); helium (0.00052 percent); methane (0.00014 percent); krypton (0.0001 percent); hydrogen (0.0005 percent); nitrous oxide (0.0005 percent); ozone (0.000007 percent); and xenon (0.000009 percent). Apart from these gases, the atmosphere contains aerosols and pollutants generated by natural events such as volcanic eruption or man-made actions, such as the use of petrol-using automobile engines.The main atmospheric gases are found in roughly the same relative proportions throughout the atmosphere up to an altitude of 80km (50 miles). This zone is known as the homosphere, while within the zone above this, the heterosphere, gases are more stratified, with concentrations of heavier gases decreasing more rapidly with altitude than those of lighter gases. The atmosphere is divided into a number of layers which reach out to a height above the surface of approximately 120km (75 miles). The lowest level, the troposphere, varies in height from about 5.5km (3.5 miles) at the North and South Poles to 16km (10 miles) at the Equator, and is the zone within which most weather occurs. The troposphere is separated from the next layer, the stratosphere, by the tropopause. The stratosphere rises from the troposphere to a height of approximately 50km (30 miles) and this is divided from the next region, the mesosphere, by another zone of transition, the stratopause. The mesosphere, which extends to a height above the surface of

Variations of temperature in the Atmosphere.

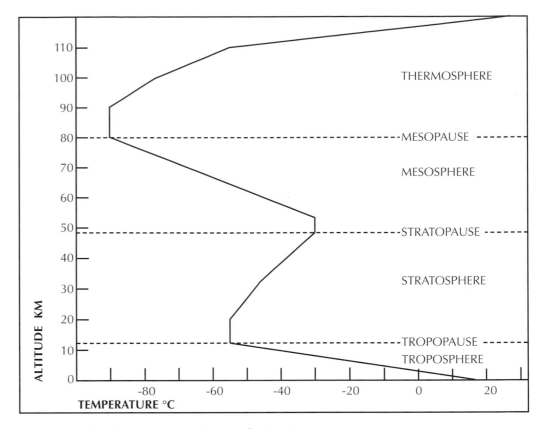

about 80km (50 miles), rises to a further zone of transition, the mesopause. Beyond the mesopause lies the outer zone of the atmosphere, known as the thermosphere. These layers are defined by the vertical variations in average air temperature that are found within each. In the case of the lowest level, temperature generally falls with altitude at a rate of 6.5°C per 1,000m (3.5°F per 1,000ft). For the last 30 years, researchers have become increasingly interested in the stratosphere. Although our weather is concentrated in the lower troposphere, it seems highly probable that climatic change is influenced by events within this zone. Little air moves between the stratosphere and troposphere and any pollutants that reach the lower regions of the stratosphere may remain in position for long periods. Gases thrown out by volcanic eruption may remain in the lower stratosphere and the ozone layer in the stratosphere is being weaken by the build-up of manmade pollutants.

See: Aerosols, Argon, Carbon Dioxide, Exosphere, Helium, Hydrogen, Krypton, Mesopause, Mesosphere, Methane, Neon, Nitrogen, Nitrous Oxide Oxygen, Ozone, Photosynthesis, Stratopause, Stratosphere, Thermosphere, Tropopause, Troposphere, Water Vapor, Xenon.

Atmospheric Pressure

The pressure (weight) at a particular point on the Earth's surface equivalent to the weight of a column of air above that point. At sea level, the pressure is 14lb per square inch. As the height above sea level increases, the column of air is shortened and the pressure per square inch decreases. Barometers are used to record changes in atmospheric pressure due to fluctuations in the weather, and millibars are the unit of measurement.

See: Atmosphere, Barometer, Millibars.

Autan

A southerly wind which blows from the Mediterranean Sea along France's Garonne river valley.

See: Wind.

Autumn

One of the four seasons of the climatic year found in temperate zones of the midlatitudes, each of which is characterized by particular conditions. In the northern hemisphere, autumn generally falls between September and November; in the southern hemisphere, autumn falls between March and May. This variation reflects the inclination of the Earth's axis and its revolution about the sun.

See: Fall, Seasons.

Backing
The anticlockwise change of direction of a wind, for example, from west to southwest to south.
See: Veering, Wind.

Ball Lightning
A glowing ball of spherically shaped air that is electrically charged.
See: Lightning.

Ballonsonde
This sounding balloon is used by meteorologist to take readings of temperature, pressure, and humidity at different height. It consists of measuring instruments taken aloft by a balloon filled with hydrogen. Finally, the balloon bursts at a height several miles above the Earth's surface, falls to the ground, and is recovered. It has been replaced to some extent by the more advanced radiosonde.
See: Radiosonde.

Badlands
Originally used to describe areas of the western United States, especially western South Dakota, this term is given to rainless uplands where the low levels of precipitation are insufficient to generate little more than sparse vegetation.
See: Arid, Drought, Precipitation.

Baguio
The name given to a tropical cyclone (typhoon) that tracks over the Philippine Islands from its point of origin in the western Pacific Ocean.
See: Tropical Cyclone, Typhoon.

Ball Lightning
A parcel of electrically charged air that has the appearance of a glowing sphere.
See: Lightning.

Banner Cloud
A type of cloud that develops on the lee side of a mountain. Air is forced to rise due to the barrier effect of the high ground, which leads the water vapor in the air to condense, thereby forming the banner cloud. Once over the mountain barrier, the cloud begins to descend, its water drops are warmed, and the banner cloud gradually disappears. As there is a constant flow of air over the mountain undergoing the same process, the presence of a banner cloud appears to be permanent, although they are constantly forming and disappearing.
See: Cloud, Condensation, Water Vapor.

Barogram
The record of atmospheric pressure made by a barograph.
See: Barograph.

Barograph
A self-recording form of the barometer in which a continuous trace of variations in atmospheric pressure is recorded on a barogram attached to a rotating drum powered by clockwork. Of the two main variants, one records the rise and fall of pressure recorded by a mercury barometer; the other, more usual, type is a modified form of the aneroid barometer and is known as the aneroidograph. Although both types are able to record the rise and fall of pressure over a set period of time, neither are particularly useful for recording pressure at a set time.
See: Aneroidograph, Barogram, Barometer.

Barometer *(Below)*
In weather forecasting, this instrument is used to measure atmospheric pressure. The mercury

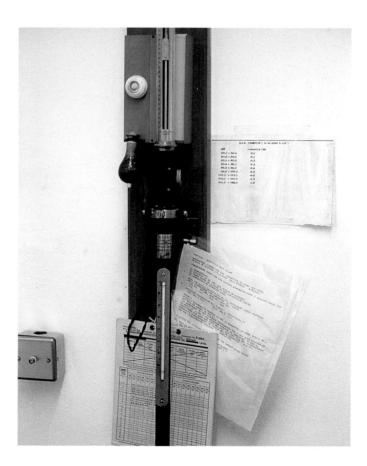

barometer, the most common and simplest form, consists of an amount of mercury held in a vacuum at the bottom of a sealed glass tube. The column of mercury is thus held in balance with the weight of the atmosphere at any given time. As the atmospheric pressure falls (the air getting denser and heavier), the column of mercury falls and as the pressure rises (the air becomes less dense and lighter) so too does the column of mercury. Atmospheric pressure is recorded by a barometer in millibars.
See: Atmospheric Pressure, Millibar.

Barometric Pressure
See: Atmospheric Pressure.

Barometric Tendency
See: Pressure Tendency.

Beaufort Scale *(Right)*
Created by Rear Admiral Sir Francis Beaufort (1774-1857), the Beaufort scale originally consisted of a series of numbers from 0 to 12, each representing the force (strength) of a particular wind and the conditions of the sea at the time. Force 0 indicates a wind speed of from 0 to 1mph and dead calm seas; Force 12 indicates a hurricane-force wind of 74mph and exceptionally high and hazardous seas. Beaufort, who served as the Hydrographer of the Royal Navy between 1829 and 1855, kept a log of wind and weather conditions at sea during his time of active service and used these to create the scale of wind strengths that bears his name and was published in the early 1800s. The Beaufort Scale became widely known and remains in use throughout the modern world in one form or another. In 1926, the scale was modified to reflect wind speeds on land, and in 1955 the US Weather Bureau added force number sfrom 13 to 17 to indicate even higher wind strengths, but as these are very rarely reached, their usage has been limited.

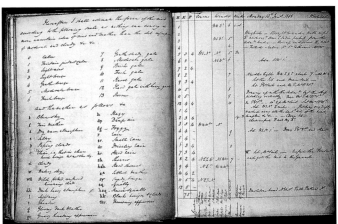

Force	Wind speed (mph)	Wind Description
0	0-1	Calm
1	1-3	Light Air
2	4-7	Light Breeze
3	8-12	Gentle Breeze
4	13-18	Moderate Breeze
5	19-24	Fresh Breeze
6	25-31	Strong Breeze
7	32-38	Strong Wind
8	39-46	Near Gale
9	47-54	Strong Gale
10	55-63	Storm
11	64-73	Violent Storm
12	74	Hurricane

See: Breeze, Hurricane, Wind.

Bergeron Process
Developed by the Scandinavian meteorologist Tor Bergeron in the early 1930s, this process explains the growth of water droplets and precipitation in cold clouds (below 0°C/32°F) in middle and high altitudes. It requires the presence of ice crystals and supercooled water droplets within a cloud. In a gradual process, the ice crystals grow as water droplets are deposited on them. As the ice crystals

grow, they collide and take up other supercooled water droplets and other ice crystals. Eventually, they become so heavy that they fall to earth. They will land as snowflakes if the air temperature is lower than freezing for most of their descent or as rain if it is higher.
See: Precipitation, Rain, Snow.

Berg Wind

Active in South Africa, particularly along the country's cool western coast, this is a warm and dry wind lasting for between two and three days at a time. It is found most regularly during winter, when the atmospheric pressure is low over the adjacent Atlantic Ocean and the country's inland plateau is the focus of a strong anticyclone. The berg wind thus blows out from the plateau and heats up to temperature often above 100°C (212°F) as it descends to lower-lying areas, especially if the wind develops in summer.
See: Anticyclone, Atmospheric Pressure.

Bioclimatology

The study of climate in relation to its impact on living organisms including humankind. Its two key objectives in relation to humans are to define those climates most suitable for human habitation and identify where in the world such climates exist.
See: Climatic Zones.

Bise

Usually accompanied by dense cloud, this cold and dry wind from the northwest, north, or northeast is found in Switzerland and southern France.
See: Wind.

Black Ice
See: Glazed Frost.

Blizzard *(Illustrated on pages 44/5, 47)*

A very strong wind that carries large amount of swirling snow and may last for several days in certain areas, particularly in the Arctic and Antarctic. Much of the snow is often previously picked up from the ground, although some snow falls from the accompanying clouds. In the northern United States, blizzards are generated in the rear of eastward moving depressions which disturb the calm of the established winter anticyclones found in the region.
See: Anticyclone, Depression, Snow, Wind.

Blocking System

Usual patterns of weather can be disrupted by the establishment of a blocking system of high or low pressure that prevents the normal movement of dominant weather systems. The effect of a blocking weather system can be severe, particularly if it is persistent. Drought or floods, and extremes of temperature are common features with such systems. In the US, the oscillation of westerly wave patterns of wind and associated weather systems can become so great that massive parcels of air become separated from the normal westerly winds and are fixed in position, blocking the normal, regular flow, thereby bring extremes of weather to the region above which they are located. Blocking systems can consist of either high or low pressure air masses.
See: Air Mass, Drought.

Blood Rain

This type of rain is tinted red because its water droplets contain dust particles that have been carried along at very high altitudes over long distances after being picked up from desert areas by wind. The phenomenon is particularly common along the Mediterranean coast of southern Europe, where dust from the Sahara desert is often deposited as blood rain, leaving easily recognizable red patches on the ground.
See: Rain, Wind.

Blossom Showers

The name given to the heavy rains that fall between March and May on the coffee-growing areas of Southeast Asia during the monsoon season.
See: Monsoon, Rain.

Bohorok

A warm and dry wind found in Sumatra during the monsoon season. It is created by the descent of an air mass on the lee side of nearby mountains.
See: Air Mass, Fohn, Monsoon, Wind.

Bora

Found in northern Italy and down the eastern margin of the Adriatic Sea, the bora, a katabatic wind, is a cold and normally dry wind, usually accompanied by clear skies, blowing from the north or northeast. Occurring primarily in winter, it is produced by high atmospheric pressure over the Balkans or Central Europe and low pressure over the Mediterranean. If a depression is centered over

the Adriatic Sea, however, the bora may be accompanied by dense cloud and rain or snow. It can last for several days and winds can reach speeds of over 90mph, though the average speed is more likely to fall between 30 and 60mph.
See: Atmospheric Pressure, Depression, Rain, Wind.

Boreal

Meaning belonging to the north. In climatic terms, boreal refers to regions, notably the coniferous forest zones of North America, Europe, and Asia between 50 and 65 degrees north latitude, with extremely cold, persistent winters and short, cool summers. They are not found in the southern hemisphere. Boreal zones have little precipitation, temperatures are uniformly low throughout the year and are marked on charts by the letter E. Two types of air mass are especially associated with boreal regions: continental polar and arctic.
See: Air Mass, Air Mass Modification, Climatic Zones.

Brave West Winds

The name given to the westerly planetary winds found in the temperate zones of the southern hemisphere. Particularly common in southern Chile, New Zealand, and Tasmania, the winds blow with great force across the open South Pacific and are accompanied by heavy rainfall.
See: Climatic Zones, Planetary Winds, Precipitation, Roaring Forties.

Breeze

An air current that is too gentle to be called a wind. On the Beaufort Scale, a light breeze (4-7mph) is categorized as Force 2 and a strong breeze (25-31mph) as Force 6.
See: Beaufort Scale, Land Breeze, Sea Breeze, Wind.

Brickfielder

The name given to a wind that blows across southeastern Australia, usually during the summer. The wind is accompanied by clouds of dust, and a prolonged spell of hot weather during which daily temperature frequently rises above 38°C (100°F). Preceding the southerly burster wind, the brickfielder is produced by the movement of tropical air into the region in front of a trough of low pressure or a depression.
See: Depression, Fahrenheit Scale, Air Pressure, Summer, Temperature, Wind.

Brockenspectre *(Illustrated page 46)*

Named after a peak in Germany's Harz mountains ("Specter of the Broken"), this is the massively magnified shadow of an individual which appears on a bank of cloud or mist in mountain areas.
See: Cloud, Mist.

Bruckner Cycle

Developed by Bruckner in the late 19th century, this cycle refers to irregular natural variations in climatic and other phenomena. In terms of climate, this includes the alternation of warm and dry years with cold and damp years. Consequences of such variations include the retreat and advance of glaciers. Bruckner estimated that the average length of a cycle was 35 years, but individual cycles could last between 25 and 50 years.
See: Climatic Change.

Buoyant

This refers to an air current that is warmer and therefore lighter than its surroundings.
See: Air Mass, Wind.

Buran

A strong northerly or northeastern wind found in Central Asia and Russia chiefly during the winter. The wind reaches gale force, carries snow and ice (when it is known as a Purga wind), and brings low temperatures up to -29°C (-20°F). The buran is active in the rear of a depression.
See: Depression, Fahrenheit, Gale, Ice, Purga, Snow.

Buys Ballots Law

A simple method of identifying the direction of movement of an area of high or low pressure. If an observer stands with his/her back to the wind in the northern hemisphere, atmospheric pressure will be lower on his/her left hand, and the opposite applies in the southern hemisphere. Consequently, in the northern hemisphere winds blow anticlockwise around areas of low pressure and clockwise around areas of high pressure. The process is reversed in the southern hemisphere. The law was named after Dutch meteorologist Christoph Buys Ballot (1817–90) who published his findings in *Comtes Rendus* (1859). Buys Ballot also invented a system of weather signals.
See: Air Pressure, Geostropic Wind.

Cacimbo

Found in the evenings and mornings along the Atlantic coast of Angola in West Africa, a Cacimbo is a heavy mist associated with low cloud. They are usually found in the dry season, when they make the air extremely moist, and tend to disappear with the dry season. A similar phenomenon, known as the "Smokes," is also found along the coast of Guinea to the north.

See: Cloud, Mist, Seasons.

Calina

This is a type of haze which reduces visibility and changes a blue sky into a dullish gray color. Produced by strong winds which draw up dust particles from the surface, it is frequently found around the Mediterranean coast between July and August.

See: Haze, Wind.

Calm

This refers to a situation in which there is virtually no horizontal movement of wind. Wind speed is usually less than 1mph and is described as Force 0 on the internationally recognized Beaufort scale.

See: Beaufort Scale, Wind.

Calms of Cancer

This is an area of calm or very light winds found within the belt of high pressure associated with the Tropic of Cancer.

See: Air Pressure, Beaufort Scale, Horse Latitudes, Tropic of Cancer, Wind.

Calms of Capricorn

This is an area of calm or very light winds found within the belt of high pressure associated with the Tropic of Capricorn.

See: Air Pressure, Beaufort Scale, Horse Latitudes, Tropic of Capricorn, Wind.

Cap Cloud

A type of cloud found directly above the summit of a mountain or high ground. The cloud appears stationary, but in reality is constantly being created on the windward edge of the peak and decaying on its leeward (downwind) side. Cap clouds have a smooth appearance and are usually associated with, and appear above, cumulus or cumulonimbus clouds. They are created when parcels of warm air (thermals) force a stable, humid layer of air above its condensation level, thus forming the cloud. Also known as Pileus.

See: Cloud, Cumulonimbus, Cumulus, Condensation, Thermal.

Carbon Dioxide

This colorless and odorless gas is found in the Earth's atmosphere. It is formed during breathing, the decomposition or burning of organic compounds, and in the chemical reaction between carbonates and acids. It accounts for 0.035 percent of the atmosphere, but is an important component. It is vital to photosynthesis and, in partnership with water vapor, causes the lower atmosphere to retain heat and make the planet more able to support life. Carbon dioxide is also recognized as a contributing factor in climatic change. Although the role of carbon dioxide in atmospheric warming was suggested in the first half of the nineteenth century when much of the world was undergoing industrialization, it has been in the twentieth century that the concentration of the gas in the atmosphere has been rising rapidly. The first noticeable increase in carbon dioxide was reported in 1938 by a British engineer, G.S. Callender. There are two chief causes for the rise. First, the burning of fossil fuels (coal, oil, and natural gas) which releases carbon dioxide, has grown enormously. Secondly, the hunger for agricultural land has increased carbon dioxide levels because of forest clearance and burning; on top of this, the loss of trees stops the process of photosynthesis by which trees remove carbon dioxide from the atmosphere. Estimates suggest that approximately 80 percent of carbon dioxide entering the atmosphere comes from the burning of fossil fuel and 20 percent from clearing forests. Approximately half of carbon dioxide released into the atmosphere remains there, while most of the remainder is taken up by the world's oceans.
See: Atmosphere, Climatic Change, Global Warming, Greenhouse Effect, Photosynthesis.

Carbon Monoxide

This gas is colorless, odorless, poisonous, and highly flammable. It is created when carbon burns in an environment with insufficient air. In industrialized countries, the chief source of carbon monoxide is the incomplete burning of fossil fuel, while it is also produced by forest clearance and burning. The gas, if present in high concentrations, can have adverse effects on health, especially on breathing. Carbon monoxide is one of the key components of smog, which is often found in urban areas under the right conditions. Its effects can sometimes be severe: 4,000 Londoners died of various illnesses and accidents made worse or caused by a particular smog incident in 1952.
See: Atmosphere, Greenhouse Effect, Smog.

Castellanus

A type of cloud usually occurring in altocumulus, cirrocumulus, cirrus, and stratocumulus formations. In appearance, Castellanus consists of a line of narrowish columns rising from a single cloud base. Castellanus indicate a high degree of instability and, if well-developed, they often herald the arrival of a thunderstorm within the next 24 hours.
See: Altocumulus, Cirrocumulus, Cirrus, Cloud, Cloud Classification, Stratocumulus, Thunderstorm.

Celsius Scale

The name often given to the Centigrade scale of temperature measurement developed by Anders Celsius (1701–44). Celsius, a Swede, devised the system in 1742.
See: Centigrade Scale, Fahrenheit Scale, Temperature.

Centigrade Scale

One of the most widely used scales for measuring temperature in the world. The system consists of two fixed positions, the freezing point of water (0°C) and its boiling point (100°C). To convert Centigrade measurements to those of the Fahrenheit scale, the Centigrade reading needs to be multiplied by 1.8 and 32 added.
See: Fahrenheit Scale, Temperature.

Centripetal Force

Meaning center-seeking, this refers to the imbalance of forces that act on a wind if it follows a curved path. It is responsible for a change of direction and not a change in speed.
See: Coriolis Force, Wind.

CFCs

See: Chlorofluorocarbons.

Chergui

This is a hot and dry easterly wind found in Morocco, North Africa. Blowing from the Sahara,

it is often responsible for drying out the ground.
See: Wind.

Chili
Found in Tunisia, North Africa, this is a dry and hot southerly wind that is much the same as the Sirocco.
See: Sirocco, Wind.

Chinook
This is a warm and dry wind that is prevalent along the eastern edge of the Rocky Mountains in both the US and Canada, and occurs during winter and spring. It is generally associated with the southern edge of a depression moving eastward across the continental land mass. It is a southwesterly wind which flows down the leeward side of mountains, but the direction in which it blows is often much modified by local physical geography. As the depression is dried and cooled as it descends after passing over the mountains, the chinook (the name means "snow eater") can raise temperatures rapidly and is responsible for melting snow and drying the land. Consequently, chinooks are of considerable economic importance. If they blow constantly, they can make the grazing of livestock possible throughout the winter months, but their absence can spell loss of grazing and livestock. A typical chinook occurs when a parcel of mild air rises to pass over a mountain range. As it descends on the downwind side, its temperature rises at the adiabatic lapse rate. The greater the descent, the larger the rise in temperature. Records indicate that the warmest chinooks can melt more than two feet of snow in a matter of hours. Chinook winds can be violent, occasionally reaching speeds of up to 100mph.
See: Adiabatic Lapse Rates, Depression, Fohn, Santa Ana, Wind, Zonda.

Chlorofluorocarbons
A group of manmade chemicals widely thought to have a significant role in the destruction of the Earth's ozone layer and, therefore, significant contributors to global warming. CFCs were first developed in the late 1920s and have many applications, including their use as cooling agents in refrigerators and air conditioning systems. They are also employed in the manufacture of insulation and are most commonly used in aerosol sprays. Because of the threat that CFCs pose to the ozone layer, their use was supposedly phased out by an international agreement in the mid-1990s, although older appliances still use them. When CFCs are released into the lower atmosphere, movements within that zone gradually transfer them to higher levels. In the stratosphere at heights above about 15 miles, intense ultraviolet radiation begins to break down the CFCs, which release chlorine gas. Chlorine reacts with the ozone, creating oxygen and reducing the thickness of the ozone layer. A thinner ozone layer allows much more ultraviolet radiation to reach the Earth's surface rather than being reflected back into space by the ozone. Consequently, a thinning ozone layer is likely to lead to a greater incidence of cancers. Early studies suggest than a reduction of the thickness of the ozone layer of three percent could lead to a 10 percent increase in skin cancers.
See: Aerosols, Atmosphere, Climatic Change, Global Warming, Greenhouse Effect, Ozone Layer, Stratosphere.

Circumpolar Complex
A band of extremely strong winds that skirts the margins of Antarctica. Because of its strength, this band cuts off the atmosphere over the continent and contains the depletion of the ozone layer that occurs there. As these winds weaken somewhat during early spring, they allow ozone-heavy air from lower latitudes to move into the Antarctic, thereby reducing the degree of ozone depletion suffered in the atmosphere above the region.
See: Ozone, Ozone Layer, Wind.

Cirrocumulus
A specific type of high-lying cloud consisting of ice crystals in the form of either small flakes or larger masses. They are often found in groups or strung out in successive lines. The clouds themselves are white or blue-white and are found rising from heights of 20,000ft. They are frequently accompanied by virga, trails of falling ice crystals or water droplets that evaporate before reaching the ground.
See: Cloud, Cloud Classification, Mackerel Sky, Virga.

Cirrostratus
A sheet-like cloud not associated with precipitation and found at heights above 20,000ft. They comprise a thin sheet of ice crystals and are often almost invisible until the sky takes on a slightly milky look. On the approach of a warm front, cir-

rostratus clouds tend to thicken and descend, turning into altostratus clouds. One sure sign of cirrostratus is the development of a slight halo around the sun.
See: Altostratus, Cloud, Cloud Classification, Front, Precipitation, Warm Front.

Cirrus

These are wispy, thread-like clouds found at heights above 20,000ft, which consist of descending ice crystals. They are usually associated with the leading edge of an advancing warm front, where moist air is forced to rise into the atmosphere's higher levels and the freezing of water takes place. If the individual threads of cirrus begin to merge into a sheet of cirrostratus, it is evidence that a depression is moving into the area. Generally cirrus clouds do not produce rainfall, as the ice crystals melt and evaporate as they descend. There are several sub-divisions of cirrus clouds. Cirrus fibratus have a more fibrous look than ordinary cirrus and are often confused with cirrostratus. Cirrus intortus are much less well defined than normal cirrus, the trails appearing tangled, while cirrus uncinus have a distinctive hook to the cloud threads. Cirrus spissatus are a dense form of ordinary cirrus clouds, appearing gray and blocking out the sun if thick enough.
See: Cirrostratus, Cloud, Cloud Classification, Depression, Front, Warm Front.

Climate

The average recorded pattern of weather at a particular place and measured over a specific, usually long, period of time. It is influenced by the relative position of oceans and continents, latitude, local conditions, altitude, and the impact of man's activities.
See: Climatic Change, Climatic Zones, Global Warming, Greenhouse Effect, Weather.

Climatic Change

The Earth's climate changes over time because of factors, several only partly understood, internal and external to the planet. These factors include: variations in the sun's activities, which affect the amount of radiation reaching the Earth. Changes in the composition of the Earth's atmosphere brought about by sulfur oxide gases thrown out by volanic eruptions, the advance and retreat of glaciers, the burning of fossil fuels, or the release of chlorofluorocarbons influence the type and intensity of radiation reaching the planet's surface. The relative sizes and positions of the Earth's oceans and continents are, through the process of continental drift, also believed to impact on climate. Some of these effects may be long term, others will probably produce much shorter-lived variations. Equally, while some may lead to an overall cooling of the planet, others are likely to raise its temperature. Scientists know that the Earth's climate will change, but are less sure about what form or intensity the change will take.
See: Atmosphere, Chlorofluorcarbons, Continental Drift, Global Warming, Greenhouse Effect.

Climatic Zones

A means of dividing the Earth into distinctive zones based on the climate occurring within a particular region. There are several chief climatic zones: boreal, continental, desert, maritime, Mediterranean, mountain, polar, subtropical, temperate, tropical humid.
See: Individual climatic zones.

Climatology

The study of the Earth's various climates and their impact on the natural environment.
See: Climate, Climatic Change.

Cloud *(Illustrations: Ns and St p49; As pp51 and 53; Ac and Ci p58; see below for abbreviations)*

A mass of small water droplets or ice flakes in the Earth's atmosphere. Clouds are formed because of the evaporation of surface water and its subsequent condensation or deposition. The clouds develop when the air carrying evaporated water begins to condense as it rises and cools or comes in contact with a cool surface. The point at which condensation begins is known as the dew point.
See: Adiabatic Lapse Rates, Cloud Classification, Condensation, Convection, Deposition, Dew Point, Instability, Inversion of Temperature, Orographic Cloud, Stability, Thermal.

Cloud Classification

Attempts formally to categorize cloud types date back to the early 1800s when Luke Howard, a British naturalist, devised a scheme based on his observations in which Latin names were given to each distinct formation. Howard's system, which is still in use today although with slight modifications, was based on two factors: appearance and cloud height above the ground. With regard to cloud appearance, there are three distinct clouds types known as cirrus, cumulus, and stratus. Cirrus have a fibrous look, cumulus are puffier, billowing clouds, and stratus have distinct layers. In terms of altitude, four distinct groups were identified: low clouds, middle clouds, high clouds, and those with a marked vertical shape, that is stretching across some of the previous divisions.

The table on page 54, based on work which appeared in *Understanding Our Atmospheric Environment* by M. Neiburger, J.G. Edinger, and W.D. Bonner (New York, W.H. Freeman, 1973), summarizes cloud classification and appearance. Heights are approximate as the altitude of a cloud's base above the Earth's surface varies with latitude.

The various names given to the various cloud types have their origins in Latin. For example, alto means high, cirro means wisp of hair, cumulus means mass, and stratus means strewn.

Types	Height of cloud base	Appearance
High Cloud		
Cirrus (Ci)	5–10 miles	Delicate streaks or patches
Cirrostratus (Cs)	5–10 miles	Transparent, thin white sheets or veils
Cirrocumulus (Cc)	5–10 miles	Layers of small white puffs or ripples
Middle Cloud		
Altostratus (As)	1.25–5 miles	Uniform white or gray sheets or layers
Altocumulus (Ac)	1.25–5 miles	White or gray puffs or waves in patches or layers
Low Cloud		
Stratocumulus (Sc)	0–1.25 miles	Patches or layers of large rolls or merged puffs
Stratus (St)	0–1.25 miles	Uniform gray layers
Nimbostratus (Ns)	0–2.5 miles	Uniform gray layers with falling precipitation
Vertical Cloud		
Cumulus (Cu)	0–1.9 miles	Detached heaps or puffs with sharp and flat bases outlines, slight or moderate vertical extent
Cumulonimbus (Cb)	0–1.9 miles	Large puffy clouds of great vertical extent with smooth or flattened tops, frequently anvil-shaped, from which showers fall, often with thunder

Low, middle, and high clouds can also be classified as stratiform clouds. They are generated by parcels of air rising very slowly, usually at a rate of less than one mile per hour, over large areas and, as they rise, they spread laterally, giving them the layered (stratiform) appearance. In the case of clouds that develop vertically, these rise much more rapidly, over 60mph on occasion, and cover a much smaller area of the sky. These are known as cumuliform clouds because of their heaped and bulbous appearance. The table on page 54 shows the four distinct groups of cloud types based on height of their base above ground and overall appearance. However, clouds can be given even more precise names based on a more detailed description of their look (see table below). Consequently, it is possible to categorize a cloud using these extra descriptions as, for example, altocumulus castellatus. This refers to a mid-altitude cloud, normally associated with white or gray puffs or waves in patches or layers, that on this particular occasion has a marked vertical development.

See: Cloud and entries on individual cloud types.

Cloudburst

An extremely heavy period of rainfall that is usually shortlived and frequently associated with a thunderstorm.

See: Cloud, Thunderstorm.

Cloud Condensation Nuclei

These are minute solid or liquid particles on which water vapor condenses to produce clouds. Some occur naturally and others are produced by man. Natural nuclei are generated by forest fires,

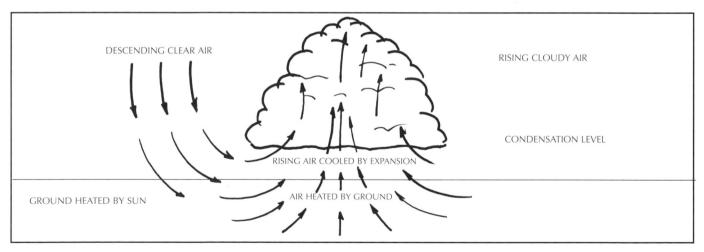

DESCENDING CLEAR AIR

RISING CLOUDY AIR

CONDENSATION LEVEL

RISING AIR COOLED BY EXPANSION

GROUND HEATED BY SUN

AIR HEATED BY GROUND

Name	Description	Cloud Type
Castellatus	Towering vertical development	Cirrocumulus, altocumulus
Congestus	Crowded in numerous heaps	Cumulus
Fractus	Fragmented	Stratus, cumulus
Humilis	Minimal vertical development	Cumulus
Lenticularis	Shaped like a lens	Altocumulus, cirrostratus, stratocumulus
Mammatus	Hanging and bulging	Cumulonimbus
Uncinus	Shaped like a hook	Cirrus

volcanic eruptions, the erosion of soil by wind, or salt water taken into the atmosphere. Manmade sources include the byproducts of the burning of fossil fuels.

See: *Chlorofluorocarbons, Cloud, Global Warming, Greenhouse Effect, Rain.*

Cloudiness

The amount of cloud cover in the sky without reference to the type of cloud visible. Generally, it is measured by eye and can be expressed in eighths, known as okta, or tenths of sky covered. For example, 8 represents total cloud cover, while 0 indicates a sky without cloud.

See: *Cloud.*

Cloud Seeding *(Hole caused by seeding illustrated page 55)*

This is a type of weather modification with the aim of inducing precipitation by mimicking the natural processes in a cloud that lead to rainfall. It is usually carried out by an aircraft which is capable of shooting flares filled with silver iodide or throwing out dry ice pellets. These substances are used to increase the rate of ice crystal growth within a cloud by encouraging water droplets there to freeze. Although cloud seeding does appear to stimulate cloud growth and precipitation, there is considerable debate over the degree to which rainfall is increased. Supporters have suggested that it can increase rainfall between 10 and 20 percent, but their opponents question these figures. They have also argued, with some justification, that inducing a cloud to produce rain over one area might well deny another area of the rain that might have fallen there if the cloud had been allowed to develop without human interference.

See: *Cloud, Precipitation.*

Cloud Street

This term refers to long lines of cumulus clouds that stretch downwind. Cloud streets develop when major thermals are long lived and occur with a steady wind speed. Several may be visible, lying parallel to each other. These develop when convection is restricted by a stable layer of air and the shear of the wind. The distance between the separate streets is usually two or three times the depth of the convection layer. If convection becomes stronger, the cloud streets are likely to develop into stratocumulus clouds.

See: *Cloud, Cloud Classification, Convection, Cumulus, Shear, Stratocumulus, Thermal.*

Col

The area between two depressions or a pair of anticyclones which face each other. The nearly uniform air pressure within the col is always higher than within the two depressions which it divides. In the case of anticyclones, the col is of lower pressure.

See: *Anticyclone, Depression.*

Cold Air Advection

Advection is the movement of a large parcel of air from one position to another. The cold air type of advection takes place when the prevalent wind blows from a cold region to a warmer one.

See: *Advection, Warm Air Advection.*

Cold Front

Located at the Earth's surface, this is the dividing line between a body of advancing cold air and a mass of warm air, below which the cold air advances in the shape of a wedge. In the northern hemisphere, cold air pushes southward, where it attempts to push past masses composed of warm air attempting to push northward. As they meet, the heavier cold air slides under the lighter warm air. The surface of a cold front rises at a much steeper angle than that associated with a warm front and, as it is traveling at a greater speed, gradually forces the warm front to rise. Typically, a cold front advances steadily at about 20mph. As the rising takes place, water vapor condenses rapidly, producing a band of swiftly rising clouds along the border between the two fronts. Clear indicators of the passage of a cold front are a drop in temperature, a veering of the wind, a rise in atmospheric pressure, and the onset of heavy showers. On certain occasions, usually when a cold front is advancing at speeds of over 25mph, a squall line develops either at the leading edge of the front or some distance ahead of it. These are associated with thunderstorms

See: *Cold Wave, Front, Occluded Front, Squall Line, Thunderstorm, Warm Front.*

Cold Wave

This is a mass of cold air, usually originating in the polar regions, found behind a cold front which is felt after the passage of a depression over a particular area. Cold waves are found regularly in North America and Siberia, where great pools of cold air build up over these extensive land masses during winter. They are much less common in the south-

ern hemisphere, as the smaller land masses there prohibit the build up of such large masses of cold air.
See: Air Mass, Cold Front, Depression.

Compressional Warming
A rise in the temperature of a specific air mass that accompanies an increase in the pressure acting upon it. The most common form of compressional warming occurs when a parcel of air descends through the atmosphere.
See: Adiabatic Lapse Rates, Atmosphere.

Condensation *(Below)*
This refers to the process by which a substance is transformed from a vapor to a liquid. With regard to cloud formation, they are formed by the condensation of water vapor in the atmosphere. Water vapor contained in the air rises as the atmosphere warms, expands, and is then cooled until the water vapor condenses out as water droplets. As the drops grow, they may fall as rain or snow, but may not reach the ground because of heating and evaporation, the opposite of condensation. The variable point in the atmosphere at which condensation begins is known as the condensation level.
See: Atmosphere, Evaporation Precipitation, Water Vapor.

Confluence Zone
The point at which streams of air, frequently but not invariably with different characteristics, meet.
See: Air Mass.

Continental Climate
One of several distinctive climatic zones defined by climatologists. The term refers to the climate found in the core areas of the great continental land masses, especially in the northern hemisphere. However, the influence of continentality spreads way beyond the cores of the great land masses to their coastal regions and offshore islands, which are exposed to bitterly cold winds from the interior during the winter. A continental climate is typified by extremes of temperature, both daily and seasonally, little rainfall, and low humidity.

Continental Drift
Developed by the German geophysicist Alfred Wegener in the early 1900s, this theory relates to the interrelated movement of the large plates that make up the Earth's surface and is now widely accepted. The plates can be oceanic or continental, or a mixture of both, and their margins move in relation to each other. Some plates are rubbing past each other, occasionally producing earthquakes if the movement is sudden and violent, others are colliding to produce mountains, others are adding material to the Earth's surface, and others are removing material. The latter two are often identifiable by volcanic activity. Evidence of the position of the continents in the geological past and the climates of the time have been uncovered by studying various remains, including fossils. Continental drift, measured over millions of years, is an example of natural climatic change.
See: Climatic Change.

Continental Glacier
A large area of ice, more usually referred to as an ice sheet.
See: Ice Sheet.

Continental Polar Air
A type of major air mass. Typically, it develops over the core areas of continents in higher latitudes, chiefly in the northern hemisphere. This is a dry wind that has distinctive characteristics which vary with the seasons. In summer, the area of origin is snow free because of long periods of

sunshine and the air is mild and dry, while in winter snow cover in the source region and cold days means that the wind is bitingly cold and dry.
See: Air Mass, Air Mass Modification.

Continental Tropical Air

A type of major air mass which develops in the core areas of continents in latitudes close to the equator. In North America, continental tropical air develops in the deserts of the southwest US and Mexico and is primarily found in the summer. It is associated with hot and dry conditions in summer and is warm and dry in winter.
See: Air Mass, Air Mass Modification.

Contrail

These condensation trails are produced by jet aircraft and develop when hot and humid exhaust fumes from the aircraft's engines mix with cold and dry air at high altitude. These manmade phenomenon may have an impact on the degree of cloud cover experienced in some areas, especially those where the number of flights are high — such as the skies over an international airport. The water vapor and cloud condensation nuclei, chiefly soot particles and tiny droplets of sulfuric acid, expelled by a jet engine can lead to the formation of larger clouds. Contrails are quickly vaporized and, by spreading in a lateral direction, form cirrus clouds. Although open to debate, it has been suggested that contrails can increase the cloud cover over a busy airport by up to 20 percent.
See: Cirrus, Cloud, Cloud Condensation Nuclei, Water Vapor.

Convection

The transmission of heat from one part of a liquid or gas to another produced by the movement

of the particles within the liquid or gas itself. Convection, essentially the vertical movement of air, is particularly important in the formation of clouds. For clouds to form in the atmosphere, the air must cool to the dew point, at which point it becomes saturated and the water vapor it contains becomes visible as water droplets or snow flakes. Air can cool by coming into contact with a cold surface, for example an ice sheet, or it can rise higher into the atmosphere. This can be achieved in four ways: it can be forced to rise over a mountain range, it can encounter a colder and denser mass of air which will force it higher, it can be squeezed upward after being confined into a restricted space, or it can rise by convection. Convection occurs when an air mass is heated from beneath. The two chief sources of this type of heat are the sun and fire. Both produce thermals, parcels of warm, rising air, which — when they reach the dew point — begin to produce water droplets, thereby forming clouds. Although the type of cloud created by convection can vary, depending on variations in the surrounding air with altitude or the stability or instability of the air, the clouds most closely associated with this process are cumuliform in nature, although altocumulus and stratocumulus types also develop.
See: Advection, Altocumulus, Cloud, Cloud Classification, Dew Point, Instability, Stability, Stratocumulus, Thermal.

Convectional Rain

A type of rainfall produced by convection processes in the atmosphere. As air warmed by the heating of the Earth's surface rises, it cools. When it reaches the dew point, water vapor condenses to form clouds. When the surface heating is great, a cloud can reach a considerable vertical depth and the water droplets can become so heavy that they fall as rain. Convectional rainfall is most common in the summer, when the surface reaches higher temperatures. Usually short but intense, summer showers can turn into thunderstorms or even hail if the convection currents are of sufficient strength.
See: Convection, Dew Point, Hail, Shower, Thunderstorm.

Convective Cloud

These are clouds of the cumulus type in which there is vertical circulation of air produced by differences in temperature within the cloud.
See: Cloud, Convection, Cumulus.

Convective Condensation Level

When convection takes place, condensation occurs at a certain point above the ground which can be gauged by the height of a cloud's base. This is the convective condensation level.
See: Convection, Convectional Rain, Convective Cloud.

Convergence

Convergence occurs when a parcel of air is forced into a confined space and some of it must rise to escape. There are increases in air pressure and density when more air converges at a point than that which escapes by rising. A typical example in the northern hemisphere is a cyclone in which surface winds blow anticlockwise and toward the center. The rising air, forced out by the inwardly driving surface air, is associated with clouds and rain.
See: Cyclone

Coriolis Force *(Illustrated on page 62)*

This is an effect generated by the rotation of the Earth about its axis and deflects major winds from their normal course. In the southern hemisphere winds are deflected to the left, while they are pushed to the right in the northern hemisphere.
See: Geostrophic Wind.

Corona

This refers to the colored ring of light that sometimes surround the sun and moon as seen through clouds at medium or high altitudes. The shape of the corona may change as the clouds pass between the viewer and subject.
See: Cloud, Cloud Classification.

Cotton Belt Climate

A type of regional climate, usually found on the eastern sections of a continent, and found in the cotton-growing areas of the US's southern states and China. It is associated with heavy rainfall, which can fall either throughout the year or chiefly in the warm season and humid conditions.
See: Climatic Zones.

Counter-Trade Winds

See: Anti-Trade Winds.

Cumuliform Cloud

This refers to clouds that cover a limited geographical area but have a great vertical height. They are effected by high rates of uplift and take on a puffy or heaped look.
See: Cloud, Cloud Classification.

Cumulonimbus

The cloud formation typically associated with thunderstorms and responsible for rain, snow, or hail. It is the largest formation associated with the cumulus family and appears as a huge mass of dense cloud that towers into the sky. The top of a cumulonimbus cloud formation can be some 15,000ft above its base, which is found approximately 6,000ft above the ground. The base regions of the cloud are usually dark gray, while its upper reaches often appear brilliant white.
See: Cloud, Cloud Classification.

Cumulus *(Illustrated pages 60 and 61)*

One of the most easily recognized cloud types. Found at low levels, they are linked to the rise of air due to convection. Typically, they look like massive balls of cotton wool. The base is near horizontal and well defined, the upper reaches are dome-shaped.
See: Cloud, Cloud Classification.

Cumulus Congestus

A type of cumulus cloud identified by its extremely vigorous growth. Although they usually consist of fairly narrow vertical towers reaching heights of nearly two miles, they can occasionally have broad bases. Rainfall associated with cumulus congestus varies, depending on the climatic zone. In temperate areas, they may herald a shower, but in tropical areas precipitation is more regular and heavier.
See: Climatic Zones, Cloud, Cloud Classification, Cumulus, Precipitation.

Current

See: Ocean Current.

Cyclone

An area of low atmospheric pressure. There are two main types: the depression, found in temperate zones, and the tropical cyclone. Although the tropical cyclone is a much more powerful and destructive phenomenon than a depression, the wind direction, if not strength, are the same in both. In the northern hemisphere, wind blows anticlockwise, while it blows clockwise in the southern hemisphere.
See: Depression, Hurricane, Typhoon. Wind.

Cyclogenesis

The term used to describe the creation and development of the low-pressure, high-wind system known as a cyclone. Cyclolysis refers to the decay of a cyclone, characterized by a rise in pressure and a slackening of wind strength.
See: Cyclone, Depression, Wind.

Cyclonic Rain

The term given to the rainfall associated with the passage of a cyclone or depression. It is generated by a warm and moist mass of air rising above a parcel of colder, denser air.
See: Cyclone, Depression, Precipitation.

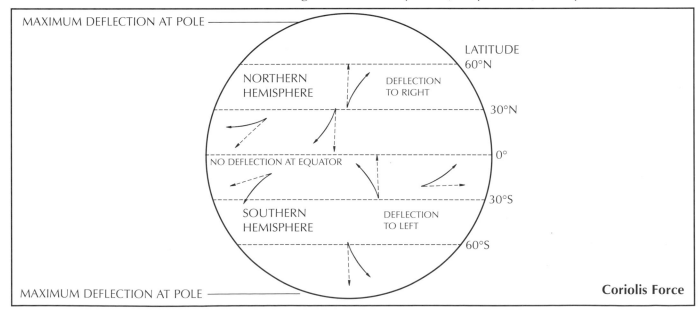

Coriolis Force

Debacle

This now little-used word refers to the breakup of winter-formed river ice during spring or summer, particularly in the former Soviet Union and North America. In general, the process of breakup takes place later in the year the farther away the ice is from the equator. Depending on position, the duration of a debacle can be two to six weeks. The melting of the ice masses leads to vast increases in the volume of water flowing in rivers and is frequently associated with extensive flooding.
See: Ice Mass.

Deepening

The name given to the process which occurs at the center of a developing cyclone or depression and produces a drop in pressure. The cause of deepening in one of these features involves the interaction of divergence and convergence within the cyclone or depression. If divergence in the higher atmosphere is greater than convergence nearer the Earth's surface, then the surface pressure at the center of the cyclone or depression deepens.
See: Convergence, Cyclone, Depression, Divergence, Filling.

Deforestation

The removal of the Earth's cover of vegetation, usually linked to humankind's exploitation of natural resources. Chief industries responsible for deforestation are logging and the clearing of land for agriculture or building settlements However, acid rain created by the burning of fossil fuels can also destroy forestry. Natural loss of cover can be brought about by volcanic eruption, although the long-term effects may be less pronounced as natural recolonization may take place. The burning of wood during clearance adds carbon dioxide to the atmosphere, thereby contributing to global warming, and the loss of vegetation also reduces the amount of photosynthesis taking place, consequently reducing the amount of carbon dioxide removed from the atmosphere. Although deforestation is common in many parts of the world, its greatest impact is on the Amazon rain forest of South America.
See: Carbon Dioxide, Global Warming, Photosynthesis.

Deposition

The process by which water vapor is transformed into ice without first being turned into a liquid. It has an important role, along with condensation, in the development of precipitation.
See: Condensation, Precipitation.

Depression

Also known as a low or cyclone, a depression is a weather feature in which the atmospheric pressure is lower than that of the surrounding air and can be identified on a weather chart by a succession of closed circular or oblong isobars. Pressure decreases the nearer to the center of the depression the isobar is. Depressions, most commonly found in the world's temperature climatic zones, develop when warm tropical air meets cold polar air, where the warm air is forced to rise over the denser cold air, thereby developing warm and cold fronts. Mature depressions can vary in extent from 100 to around 2,000 miles across and can move at speeds of up to 650 miles per day. Depressions are linked to strong winds, with the strength depending on the steepness of the pressure gradient. In the northern hemisphere, winds blow in an anticlockwise direction into the lower pressure zone at the center; in the southern hemisphere, they blow toward the center in a clockwise direction. The weather associated with depression is unsettled, and cloudy, and the likelihood of precipitation is high due to the condensation of the water vapor contained in the rising warm air.
See: Air Mass, Cold Front, Climatic Zones, Condensation, Cyclone, Isobar, Front, Occlusion, Precipitation, Pressure Gradient, Warm Front, Water Vapor.

Desert Climate *(Illustrated pp 64/5 and 66, Namib desert)*

One of the Earth's major climatic zones. A desert climate is typified by a lack of precipitation. Two distinct types have been identified: the tropical (hot) desert and the midlatitude desert. Although the total amount of rainfall (usually less than 10in per year) is important in identifying a desert climate, its distribution throughout the year can also be important. Some desert areas have annual rainfall totals above 10in, but because it falls intensively, rather than being spread throughout the year, it is lost quickly because of evaporation and surface water runoff. Most deserts are found within the interiors of large continents where moisture heavy-maritime air masses are unable to penetrate. Deserts are also usually linked to areas of pronounced and prolonged high pressure

which work against the generation of rainfall. Indeed, most of the world's great deserts, including the Sahara, are linked to zones of persistent high pressure.
See: Air Mass, Climatic Zones, Desertification, Precipitation.

Desertification *(Right)*

The process by which agricultural land is transformed into desert through the interaction of poor land management, particularly overgrazing leading to the removal of vegetation and deforestation, and adverse climatic conditions or climatic change. The extent of a desert can show annual variation, but it is prolonged advances that have the most profound impact on regional and global climates.
See: Deforestation, Global Warming.

Dew *(Below right)*

This is a product of night cooling, when an object — a rock, for example — at the Earth's surface gives up heat that it has gained from the sun or a manmade source during the day. However, this heat loss is sometimes compensated for by atmospheric radiation (heat) which is taken up by the object. However, on a clear night the overall balance of this heat loss and gain of the object favors the former rather than the latter. As the object in this case is cooler than the surrounding air, then heat in the air passes to the object and, if the cooling of the air persists, then the air in immediate proximity to the object becomes saturated with moisture. Dew will be deposited if the air temperature remains above freezing as water vapor condenses on the object. Dew is not a form of precipitation as it does not fall from clouds.
See: Condensation, Dew Point, Frost, Precipitation, Water Vapor.

Dew Point

This is the temperature at which air in the atmosphere must be cooled to become saturated with water vapor which, through condensation, is transformed into water droplets or ice flakes depending on the extent of the temperature drop. If the water vapor contained in a parcel of air is to be deposited on a surface as dew, the temperature of the surface must fall below the dew point.
See: Cloud, Condensation, Dew.

Diathermacy

The ability to transmit heat through radiation of a body or surface, such as the atmosphere or a building.
See: Atmosphere.

Diurnal Range

A term referring to the difference between the maximum and minimum of a meteorological reading, for example, pressure and temperature, over a period of 24 hours.
See: Temperature.

Divergence

A process by which surface air pressure can vary due to changes in wind speed and direction. Divergence at the surface occurs when horizontal winds blow away from a particular point. At the center, if the volume of divergent air flowing away is greater than the replacement air descending from above, then both air pressure and density

decline. The opposite of divergence is convergence, and both are responsible for changes in weather systems such as depressions and anticyclones.
See: Anticyclone, Convergence, Depression, Wind.

Doldrums

The belt of low atmospheric pressure around the equator where northeast and southeast trade winds meet to produce a broad band of east to west wind. The area is characterized by calm or light winds at the surface. Nevertheless, the doldrums are also noted by rapidly rising air and are frequently associated with thunderstorms, squalls, or heavy rain. The extent and location of the doldrums are neither constant nor fixed in position. Generally, they move either northward or southward in the wake of the sun. The most violent weather associated with the doldrums tends to be generated along the intertropical convergence zone.
See: Intertropical Convergence Zone, Trade Winds, Wind.

Downburst

A feature linked to extreme thunderstorms, a downburst is a powerful downdraft of air which strikes the Earth's surface violently and is then forced to travel horizontally as a powerful wind, capable of uprooting trees or damaging buildings. The term was first coined from study of an event in 1974 and downbursts can be categorized as either micro- or macrobursts depending on the extent of the area damaged.
See: Thunderstorm, Wind.

Drizzle

A form of unfrozen precipitation. The rain's water droplets are extremely small, measuring less than 0.02in in diameter. Weather reports recognize three types of drizzle, each reflects the impact the precipitation has on visibility. In light drizzle, objects are visible at distances of over 1,100 yards, medium drizzle restricts visibility to between 550 and 1,100 yards, and heavy drizzle restricts visibility to below 550 yards. Drizzle usually falls from stratus and stratocumulus clouds. Precipitation falling in water droplets larger than 0.02in is classified as rain.
See: Cloud, Cloud Classification, Precipitation, Stratocumulus, Stratus.

Drought

A prolonged period of dry weather brought about by an excessive removal of water from the land by evaporation or human usage over water gain from precipitation.
See: Arid, Absolute Drought, Dry Spell, Partial Drought, Precipitation.

Dropwindsonde

A device for gathering meteorological data. It consists of a pack of recording instruments fitted with a radio transmitter which is dropped from an aircraft and records vertical variation in features such as temperature and pressure.
See: Temperature.

Dry Deposition

A method by which particles suspended in the atmosphere are removed by the effect of gravity or by the particles hitting objects such as buildings on the Earth's surface. It is a natural means of removing pollutants from the atmosphere.
See: Aerosols, Atmosphere, Impaction, Gravitational Settling, Air Pollution, Scavenging.

Dry Line

The dry line, often associated with intense thunderstorms, is the dividing line between two distinctive air masses, one comprising warm, humid air and the other consisting of warm, dry air, found in the southeast zone of a well-developed midlatitude cyclone.
See: Air Mass, Cyclone, Thunderstorm.

Dry Spell

Although the term is defined in Great Britain as a period of 15 days in which the daily total of rain is less than one millimeter, this definition is not recognized internationally. In many cases acknowledgment or recognition of a dry spell depends upon the effect of a lack of rainfall on plant life rather than a set of weather readings.
See: Drought, Precipitation, Wet Spell.

Dust (Right)

Minute particles of solid matter found throughout the atmosphere, which can be transported over vast distances from their point of origin by the wind. Dust can be created by human activities, especially as a byproduct of the burning of fossil fuels, or can occur naturally, as in the case of dry

soil particles blown aloft in desert areas by dust-storms or material thrown out during volcanic eruptions. Particles can act as condensation nuclei in the atmosphere, a key component of cloud formation and rainfall.
See: Atmosphere, Cloud Condensation Nuclei, Cloud Formation, Duststorm, Precipitation.

Dust Bowl

A type of desert created by human activity. Dust bowls are usually found in areas of low or highly variable rainfall, where agricultural mismanagement is practiced. The soil is dry, the soil-binding effects of tree root systems is limited, and intensive agriculture or overgrazing by domesticated animals can further damage the soil structure. High winds remove the top soil and sudden heavy rainfalls intensify soil erosion. The term is chiefly associated with the Midwest region of the US in the 1930s, although similar events have taken place in the core of other continents.
See: Duststorm.

Dust Devil

A localized, shortlived, and smaller version of a duststorm. Dust Devils are typically no more than a few yards in diameter and develop when soil particles, blowing around the center of the storm, are taken up into the atmosphere, reaching heights of up to a few thousand feet. These storms moved at average speeds of between five and 15mph, although speeds of over 70mph have been recorded. They are generated by excessive convectional heating by the sun, usually in arid or desert areas, which creates hot spots in which rising warm air is replaced at the surface by converging surface winds. As the process develops, surface dust is lifted off the ground. Several may be generated at the same time.
See: Convection, Duststorm.

Dust Dome

A feature associated with the generally higher temperatures found in cities, usually because of the concentrated burning of fossil fuels, the thermal properties of buildings which allow heating to be transferred more rapidly into an urban atmosphere, and the lack of moisture in the atmosphere due to water being removed by sewer systems. Consequently, the higher urban temperatures compared with the surrounding countryside, lead to the development of convectional winds blow-ing into a city as rising hot air is replaced by denser and cooler air from rural areas. Rising hot urban air contains aerosols in concentrations far higher than in rural areas, and create the dust dome.
See: Aerosols, Convection, Dust Plume.

Dust Plume

Often associated with the dust domes which develop over urban areas, dust plumes are created when local winds reach a certain speed, usually 10mph, and spread a dust plume with its pollutants downwind into surrounding areas.
See: Dust Dome, Pollution.

Duststorm *(Below)*

A strong storm in which large volumes of dust particles are carried into the atmosphere and greatly reduce visibility. Such storms are generated by high, turbulent winds blowing over an arid, often desert, surface and drawing soil particles aloft in vast quantities, producing a wall of dust reaching heights of over 10,000ft. Winds in front of an advancing duststorm are usually light but rise rapidly in speed as the storm approaches. Duststorms vary in duration. If they are accompanied by thundery rainfall, they can be shortlived as the rain washes the dust out of the atmosphere.
See: Dust, Haboob, Thunderstorm, Wind.

Entrainment

A process, usually associated although not exclusively so, with well-developed thunderstorms in which unsaturated air at the margins of a cloud is drawn into that cloud. The unsaturated air mixes with the saturated air within the cloud, thereby causing some of the cloud's water droplets or ice crystals to vaporize.
See: Thunderstorm.

El Nino

This term refers to a prolonged period of unusually high sea surface temperature over the eastern tropical and equatorial Pacific Ocean, along the coast of North and South America. The phenomenon can persist for anything from a month to a year or more, and occurs about once every three to seven years. The name, given by Peruvian fishermen who noted the poor fishing associated with the high ocean temperatures, means literally "little boy" but in fact refers to Jesus Christ. The cause of this unusually high ocean temperature is not fully understood but is thought to be caused by interaction between the ocean and atmosphere. The onset of El Nino seems to follow changes in the pressure gradient between the western and eastern tropical zones of the Pacific. If the air pressure over the eastern tropical Pacific descends relative to that in the western tropical Pacific, then the trade winds begin to slacken or, if the El Nino is of sufficient intensity, reverse direction. The changes in the intensity or direction of the trade winds impact on the movement of ocean currents and the ocean's surface temperatures. The warm water currents that under the normal pressure systems travel toward the western Pacific reverse direction, heading eastward. El Nino has considerable impact on the weather in the tropics and subtropics, though the influence is rarely consistent. For example, in Hawaii El Nino is usually a bringer of drought, while, on occasion, it can bring heavy rainfall to southern California.
See: Climatic Zones, La Nina.

Equator

An imaginary line drawn around the surface of the Earth at the midpoint between the North and South Poles. It is the point of zero degrees latitude where the Earth has its greatest circumference, and marks the dividing line between the northern and southern hemispheres.
See: Heat Equator.

Equatorial Climate

This climatic zone lies between five and 10 degrees latitude either side of the equator and is most associated with low-lying areas. The climate is typically hot and humid throughout the year, with little seasonal variation. There is no clearly identifiable dry season and rainfall, generated by convection and frequently a byproduct of thunderstorms, is distributed fairly equally throughout the year. The Amazon, Indonesia, and the Philippines all have equatorial climates.
See: Climatic Zones, Convection, Convectional Rain, Thunderstorm.

Equatorial Forest *(Illustrated page 72, Puerto Rico)*

More commonly known as rain forests, these are humid and hot evergreen forests located in the world's equatorial climatic zones. Rainfall is extremely heavy and there is no dry season. They play an important part in cleaning the Earth's atmosphere by removing carbon dioxide from the air and adding oxygen during the process of photosynthesis. However, they are under threat from humankind's seemingly insatiable demand for wood and agricultural land. Burning the equatorial forests has an impact on atmospheric pollution and global warming due to the release of carbon dioxide into the atmosphere.
See: Carbon Dioxide, Climatic Change, Deforestation, Global Warming, Pollution.

Equinox

A date when every part of the Earth has 12 hours of day and an equal amount of night. There are two: the spring or vernal equinox occurs in late March, while the autumnal equinox occurs in late September.

Equipluve

A line which joins places having the same mean amount of precipitation over the same given period expressed as a percentage of the amount of rainfall that would fall if equally distributed throughout the year.
See: Precipitation.

Etesian Winds

The name given to the constant and strong northerly winds found in the eastern Mediterranean. Typically, they occur from May to October, can reach speeds of over 40mph, and blow around the large trough of low pressure

which stretches from northwest India to the Middle East. These dry winds tend to gather strength during the day, moderate during the night, and frequently carry large amounts of dust. The winds are also known by the name "meltemi." *See: Wind.*

Eustatic Movement

A significant change in sea level most associated with the advance or retreat of glaciers and ice sheets and therefore closely related to long-term climatic change. As temperatures rise over a prolonged period, ice masses melt, thereby adding to the world's oceans, which rise. As temperatures fall, ice masses increase in extent and sea levels fall. However, eustatic movement is complicated by the effects that ice sheets have on the land. Although sea levels may fall as water in the form of ice is taken from the oceans, the weight of the massive glaciers and ice sheets may lower the land and reduce the overall drop in sea level, or the land may rise as the glaciers retreat.

Evaporation

The term used to define the process by which a substance changes from a liquid into a vapor. On the Earth's surface water from the land, rivers, lakes, and oceans undergoes this process due to the warming produced by the Sun's heat. Evaporation is a virtually constant process as the air is rarely saturated with water vapor and its rate is primarily determined by the temperature of the air, the amount of water vapor already present in the atmosphere, the type of water surface, and the strength of the wind. Evaporation rates are highest in areas of high temperature in which the air is unsaturated.
See: Cloud, Condensation, Evapotranspiration, Hydrological Cycle.

Evapotranspiration *(Right)*

A component of the global hydrological cycle in which the Earth water is transferred between the atmosphere, the land, and the world's oceans. Evapotranspiration has two components: the evaporation of water from the land due to the creation of atmospheric water vapor by the Sun's heat, and transpiration, the process by which plants take up moisture from the soil through their roots which is then evaporated through their leaves.
See: Evaporation, Hydrological Cycle.

Exosphere

This refers to the outermost layer of the Earth's atmosphere. Extending for a distance of approximately 35 miles, it begins some 400 miles above the planet's surface.
See: Atmosphere.

Expansional Cooling

The chief means by which clouds form, this refers to the process in which air cools and its relative humidity increases. As air cools, it expands and moves against the pocket of air already occupying that space, which involves the use of energy taken from the heat contained in the rising air.
See: Cloud, Compressional Warming.

Eye of the Hurricane *(Below)*
The central area of a hurricane which is lacking cloud, has moderate to light winds, and in which air is descending to the Earth's surface.
See: Hurricane, Eye Wall.

Eye Wall
The circle of cumulonimbus clouds that surrounds the central eye of a well-developed hurricane.
See: Cumulonimbus, Eye of the Hurricane, Hurricane.

Fahrenheit Scale

The temperature scale devised by German physicist Gabriel Daniel Fahrenheit (1686–1736) in 1714 which became standard in many parts of the world. The system has two fixed points: freezing point (32°F) and boiling point (212°F). To convert temperatures into measures on the Centigrade scale, subtract 32 from the Fahrenheit reading and then divide the result by 1.8.
See: Celsius Scale, Centigrade Scale, Temperature.

Fall *(Illustrated pages 76/7)*

The name given to autumn in the United States, reflecting the fall of leaves that occurs during the season.
See: Autumn, Seasons.

False Cirrus

This is a cloud type that is found at the upper margins of a cumulonimbus cloud. Over time, this upper part may become detached from the cumulonimbus and has the appearance of a dense cirrus cloud
See: Cirrus, Cloud, Cloud Classification, Cumulonimbus.

Fata Morgana

The name given to a particular type of mirage. In a Fata Morgana, objects appeared to be vertically elongated because they are viewed through a number of horizontal layers of air with different degrees of refraction. These types of mirage are often found over water and are particularly common in the Strait of Messina separating the Italian mainland from Sicily.
See: Mirage, Refraction.

Filling

Also known as cyclolysis, filling refers to the decay of a cyclone, a process accompanied by a drop in wind speed and an increase of the pressure at the center of the weather feature.
See: Cyclogenesis, Cyclone, Deepening.

Flaschenblitz

A type of lightning which strikes upward from the top part of cumulonimbus clouds.
See: Cloud, Cumulonimbus, Lightning.

Flash Flood

A sudden and violent flood produced by the torrential rainfall often linked to thunderstorms. The severity of a flash flood is not only influenced by the volume or duration of precipitation, but is also conditioned by local geography. In rural, mountainous areas, heavy rainfall is often so great that the ground's ability to absorb it through infiltration is exceeded and the surface water rapidly runs off into a river channel, where the water level rises and then breaks the river bank. Flash floods can be found in urban areas, where concrete and asphalt prevent water soaking into the ground and sewer systems speed water into a river channel. When the ability of sewers to remove water is exceeded, water builds up on the impervious surface, causing localized flooding.
See: Thunderstorm.

Fog *(Below)*

Any cloud located in the lower atmosphere at the Earth's surface that reduces visibility to less than 0.6 of a mile. Fog consists of a thick mass of tiny water droplets, or smoke and dust particles, and is distinguished from mist by the size and density of the water droplets in the cloud. For weather forecasting purposes, any cloud in which visibility is more than 0.6 of a mile is considered mist. The formation of fog can begin with the condensation of water vapor in the lower atmosphere which begins when the air cools below the dew point. This process of cooling is often due to radiation cooling at the surface, particularly at night if there are light winds and a cloudless sky. Under these conditions, the fog is thickest after sunrise and is likely to disappear in the early afternoon as the sun warms the Earth. In winter, lower daytime and

nightime temperatures may encourage a fog to last for several days. Other methods of generating fog involve the movement of a parcel of air over a cold surface or the mixing of a warmer air mass with a cooler one through advection.

Fogs can occur both on the land and at sea, with the former being most common in autumn and winter and the latter being dominant in spring and summer. Sea fog is considered to be a type of advection fog and is usually associated with the major cold ocean currents. The effects of humankind's burning of fossil fuels can lead to a smoke fog, which is a mixture of smoke emissions and water droplets. This smoke is usually dense and persistent, and can lead to health problems
See: Advection, Advection Fog, Air Mass, Condensation, Haze, Mist, Radiation, Sea Fog, Smog, Smoke Fog, Water Vapor

Fog Bow

This is a white feature, similar to a rainbow, which is found opposite the sun in the presence of a fog. Unlike a rainbow, the colors overlap, thereby producing the white effect. However, the edges of a fog bow are slightly coloured. There is a reddish hue on the outer edge and a bluish tinge on the inner.
See: Fog, Rainbow.

Fog Dispersal

The presence of thick, persistent fog can be hazardous to various types of transport, especially aircraft, and several methods have been used to disperse fog, although with only limited success. One method, now considered too expensive to be really practical, is the use of heat-burners to raise the air temperature over a runway. This lowers the relative humidity of the air to below the point of saturation so the fog droplets return to vapor.
See: Fog.

Fohn

Originally the name given to the type of warm and dry wind which blows down the leeward side of mountains into the valleys of the northern Alps, but now used to describe any such feature worldwide, fohn winds are associated with the movement of a depression. In the case of the Alps, a depression moving toward the north of the mountains sucks in air from the south, which rises up the southern slopes, forming clouds that produce heavy rain. The temperature of the rising

moisture-carrying air falls as the parcel of air rises, which leads the water vapor it contains to condense, thereby producing cloud and frequently depositing snow or rain on the windward (southern) mountain slopes. However, when this parcel of air reaches the northern mountain slopes, it is still warm but has given up most of its moisture through precipitation. As it begins to descend, it heats up rapidly because of compressional warming and begins to blow down the valleys with considerable force. This is the warm fohn wind of the Alps, and it is responsible for melting snow cover in the region during spring and also has a beneficial impact on crops in autumn by helping to prolong the growing season. The direction in which the Alpine fohn wind blows is influenced by the direction of mountain valleys, but as it is a southerly wind, its impact is more intensely felt in valleys running from south to north. Fohn-type winds, ones which are transformed from being damp and cool into being warm and dry by mountain barriers, are found throughout the world and have often been given names by local people who have had long experience of them. The Chinook of the western United States along the line of the Rocky Mountains is one example.
See: Adiabatic Lapse Rate, Chinook, Compressional Warming, Depression, Orographic Cloud, Samoon.

Forced Convection

A feature usually associated with the development of thunderstorms, forced convection relates to the vertical movement of a parcel of air caused by convection, the process by which the sun's warming of the Earth surface forces air to rise, but strengthened by the presence of a front or high ground. It can also occur when surface winds converge at a particular point.
See: Convection, Free Convection, Front, Thunderstorm.

Free Convection

See: Convection.

Freezing Nuclei *(Illustrated pages 78/9)*

These are solid and liquid particles in the atmosphere which promote ice formation. Ice-forming nuclei are of two types and are less common than cloud condensation nuclei, only becoming active when temperatures fall well below freezing. Freezing nuclei are those which cause water droplets to freeze, while deposition nuclei are those that allow water droplets to be directly deposited as ice.
See: Cloud Condensation Nuclei, Deposition.

Freezing Rain

This is a form of precipitation that occurs when rain from a mild air layer descends through a parcel of air which has a temperature below freezing. The rain itself then cools dramatically as it passes through the freezing air and becomes supercooled. When it contacts a cold surface on the ground, it freezes immediately, leaving a coating of ice. If the water droplets falling as rain are smaller than 0.02in in diameter, then the phenomenon is known as freezing drizzle.
See: Drizzle, Precipitation.

Friagem

A cold wave found in the tropical grasslands of Brazil lying to the south of the Amazon basin in winter. A feature associated with the development of anticyclones, it is a period of cold weather.
See: Anticyclone, Cold Wave.

Friction

The presence of manmade and natural obstacles on the Earth's surface, such as buildings, fences, mountains, and trees, can influence the flow and speed of low-level wind. These frictional barriers set up eddies in the wind flows on the downwind side of the obstacle, slowing it down, and forcing it to deposit any material, such as snow or dust, that it is carrying. As these obstacles lie close to the surface, their impact diminishes with height and wind speed is therefore generally higher at altitude.
See: Friction Layer, Wind.

Friction Layer

The height of the zone above the Earth's surface where the effects of friction-generating obstacles no longer impinge on the flow and speed of wind, generally considered to be around 0.6 of a mile.
See: Friction, Wind.

Front

A narrow zone of transition on the Earth's surface, separating cold and warm air masses, that is created when air masses take on the characteristics (temperature and humidity) of their source region and then move into contact. At this point,

the denser cool air forces the warmer air to rise. Fronts are defined by the temperature that follows behind their leading edges relative to the temperature found ahead of the front. Consequently, a cold front means that the temperature ahead of the front is warmer than that behind it; the opposite indicates a warm front. The ascent of the warm air mass is usually sufficient for the air to cool, when it produces clouds and precipitation. Fronts are continuously forming and decaying. Fronts form when the difference between two air masses is particularly striking and they decay when the distinctive characteristics of the two air masses declines. Fronts are a major influence of weather in the world's midlatitude regions, including North America and Europe.
See: Air Mass, Cold Front, Cloud, Depression, Frontogenesis, Occluded Front, Precipitation Stationary Front, Warm Front.

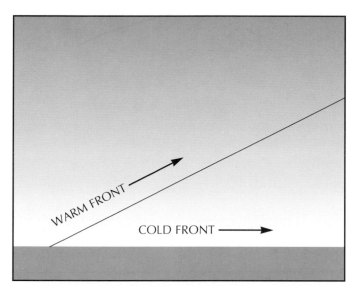

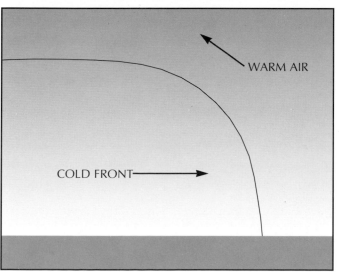

Frontogenesis
This refers to the creation and development of a front.
See: Front.

Frost
At night, the Earth's surface features give off radiation into the atmosphere as they cool and this is offset to a degree by radiation from the atmosphere, which has a heating effect on surface objects. However, if an object sends out more radiation than it receives from the atmosphere, its surface becomes cooler than the surrounding air and heat is conducted from the air to the object. If this is of a sufficient amount, the air in contact with the surface of the object becomes saturated with water. If the temperature of the air then drops below freezing, water vapor in the air immediately in contact with the object is deposited on it as frost.
See: Dew, Freezing Rain, Hoar Frost, Frost Hollow Radiation.

Frost Hollow
Also known as a frost pocket, this is an depression in the ground into which a parcel of cold air will sink from an adjacent slope due to the effects of gravity and then build up, thereby making the likelihood of severer and more frequent frosts greater. The process of frost development begins when night-time cooling of a humid parcel of air leads it to become highly saturated with moisture, which is then deposited as frost on the surface due to deposition.
See: Deposition, Frost.

Frost Point
Water vapor in the atmosphere will be deposited on a surface if the temperature of the surface falls below the frost point.
See: Frost.

Funnel Cloud
This is a swirling, tapering cloud of water droplets which descends from the base of a storm cloud, often to ground level. Funnel cloud is common in a tornado, and forms when humid air cools and expands as it is drawn towards the center of the tornado due to the steep pressure gradient that runs from the systems outer edge to its center.
See: Tornado.

Gale

A strong wind with a ground speed of 47–54mph which is defined on the Beaufort Scale as Force 8. A near gale, a wind of 39–46mph, is known as Force 7 on the same scale. The structural effect of a gale on property is usually slight, although aerials, tiles, and chimney pots may be brought down.
See: Beaufort Scale, Wind.

Garua

Found in western Peru, South America, during winter, this is a drizzle or thick mist which covers much of the region's coasts in thick cloud and generates a significant proportion of the area's limited rainfall.
See: Cloud, Drizzle, Mist, Winter.

Geostropic Wind

This is a horizontal wind which blows in a straight path at high altitudes above the Earth's friction layer. The wind's direction is generated by the interaction between two forces. One is the deflecting force that is produced by the rotation of the Earth known as the Coriolis Force. This acts on the basis of Ferrel's Law which states that a body, including a large air mass, moving in any direction across the surface of the Earth will be deflected by the Earth's rotation to the right in the northern hemisphere and to the left in the southern hemisphere. For example, if a wind were blowing from the north to the south in the northern hemisphere it would be deflected to the east (right), and if it were blowing from south to north in the southern hemisphere, it would be deflected to the west (left). The second influence on the geostropic wind is the horizontal pressure gradient. According to a law developed by Buys Ballot, winds in the northern hemisphere travel clockwise around areas of high pressure and anticlockwise around lows, while winds in the southern hemisphere blow in the reverse directions around areas of high or low pressure. These two forces, the Coriolis force and pressure gradient, eventually reach a balance and the wind flows in a straight line parallel to the isobars at a constant speed.
See: Buys Ballots Law, Coriolis Force, Friction, Friction Layer, Isobar Pressure Gradient.

Gibli

A localized hot and dry southerly wind found in Libya, North Africa, of the Sirocco type.
See: Sirocco, Wind.

Glacier Breeze *(Illustrated pp 86/7, Tschingelhorn Glacier)*
A cool katabatic wind which blows along the direction of a glacier's flow, produced by the cooling effect of air passing over the glacial ice.
See: Ice, Katabatic Wind, Wind.

Glazed Frost
An often thick covering of ice produced on physical features through the process of rain falling through an atmosphere that has a temperature below freezing point. A second method of formation occurs when a warm, moist wind follows a cold spell. The moisture condenses and then freezes on exposed surfaces. One of the more common and hazardous forms of glazed frost forms on highways and is known as black ice.
See: Condensation, Ice, Freezing Point, Freezing Rain, Wind.

Global Warming *(Illustrated page 82)*
The question as to whether the Earth is warming at a potentially dangerous rate is a matter of scientific debate. The Earth has warmed and cooled in the past as the various ice ages indicate, but the current debate focuses on the nature of humankind's impact on global warming. There is general agreement that warming has taken place since the beginning of industrialization and that the rate of warming is probably increasing as the Earth becomes more developed. Evidence for this is the accelerated retreat of glaciers, abnormal rises in sea level, the spread of desert areas, rise in the temperature of soils, and the increase in the prevalence of previously rare climatic extremes such as drought and flood. However, a note of caution needs to be sounded. No one full understands the complex mechanics and interactions of the planet's climate and humankind's influence on it. Nor can anyone state conclusively that global warming will continue at its present rate. It is also arguable that seemingly abnormal warming may be the product of other factors, including major volcanic eruptions. Many scientists argue that just because they have incomplete knowledge of the subject, it would nevertheless be neglectful not to control those human activities that in all probability contribute to global warming.
See: Climatic Change, Greenhouse Effect.

Glory
A ring of light found around a Brockenspectre.
See: Brockenspectre.

Gradient Wind
This is a extensive, high-altitude wind which blows horizontally and is unaffected by the Earth's friction layer. The wind follows a curved path and usually develops around an area of low pressure (a cyclone) or an area of high pressure (an anticyclone). In the northern hemisphere, a gradient wind blows anticlockwise around a cyclone parallel to the isobars, while it follows a clockwise path parallel to the isobars around an anticyclone. The reverse is also true in both cases in the southern hemisphere.
See: Anticyclone, Coriolis Force, Cyclone, Friction Layer, Isobar.

Grape Belt
A belt of land some 60 miles long which runs along the south shoreline of Lake Erie in the US and is typified by a long and mild autumn/fall, late frosts, and less severe winters than the other more inland regions of the US on a similar latitude due to the lake's climatic influences.
See: Autumn, Fall, Frost, Latitude, Winter.

Grassland
Found in areas where average rainfall is too low to permit the growth of forest, these regions consist of extensive grass and are found in many parts of the world. Two distinct types have been identified on the basis of the climatic zones in which they are located: tropical grasslands, of which the most well known is the savanna regions of sub-Saharan Africa, and the temperate grasslands of the North American prairies, the Asian steppes, and the South American pampas.
See: Climatic Zones.

Graupel
A German word which describes hail that is soft and slightly melted.
See: Hail.

Gravity
The atmosphere of the Earth and the movement of air within it are acted upon by the effects of gravity. Gravity has two components, the greater of which is the power of attraction between the Earth and other objects. This is known as gravitation. The second, and weaker, component is the centripetal forces given to all objects because of the action of the Earth's rotation on them. The force of gravity produced by these two factors is defined as

the acceleration of a unit mass of an object downward at a rate of 9.8m (32ft) per second per second. Gravity affects only winds that are rising or falling and, unlike other forces such as the Coriolis Force or the friction layer, does not effect the horizontal movement of wind.
See: Coriolis Force, Friction Layer, Wind.

Gravitational Settling

The process by which substances, such as aerosols in the Earth's atmosphere, descend to the surface under the influence of gravity. Generally, the larger the substance the more likely it is to settle nearer its source. Smaller particles take longer to fall to Earth and may be blown considerable distance from their point of origin by the wind.
See: Dry Deposition, Gravity, Impaction.

Greenhouse Effect *(Below)*

A feature of the industrialized world that has become of increasing concern, the greenhouse effect is generally understood to be central to the issue of global warming and a potentially dangerous climatic change. The greenhouse effect, which is also a naturally occurring phenomenon, is a product of the interaction between the Earth's surface and atmosphere and ultraviolet solar radiation. Some 31 percent of the sun's radiation

is reflected back into space by the Earth's atmosphere, while the remaining 69 percent passes through the atmosphere, where a further 23 percent is absorbed. The remaining 46 percent of the incoming solar radiation reaches the Earth surface, where the vast majority, some 71 percent, is absorbed by the world's oceans. If this absorption was to continue unchecked, then both the Earth's land and sea temperatures would rise continuously. However, this does not happen under normal circumstances, as heat in the form of infrared radiation escapes back into space from the Earth's surface via its atmosphere. A percentage of the infrared radiation emitted by the surface, however, is radiated back to the surface by the atmosphere, as it is much slower at transmitting infrared radiation back into space. This reduces the rate at which heat is returned to space and increases the temperature in the lower atmosphere. The rate at which the atmosphere reflects some infrared radiation back to the surface is determined by the volume and reflectivity of certain naturally occurring atmospheric gases. Among these greenhouse gases are water vapor, carbon dioxide, methane, nitrous dioxide, and ozone. The concern over greenhouse gases has grown in the present century as it has become clear that the amount of these greenhouse gases in the atmosphere has been rising at an increasing rate since the onset of industrialization in the middle of the 1800s. In particular, the burning of fossil fuels and forest clearance has added significant amounts of carbon dioxide to the atmosphere, thereby increasing the amount of infrared radiation reflected back to the surface and reducing the amount escaping into space. Similarly, the amount of methane in the atmosphere appears to be rising due to the dumping and decay of organic matter in landfill sites and the production of the gas in intensive agriculture and the cultivation of rice. Of more recent concern to scientists is the production of chlorofluorocarbons which damage the ozone layer, thereby allowing much more ultraviolet solar radiation to pass into the atmosphere at the outset.

The degree and impact of global warming is a matter of debate. Some commentators believe that, if it is allowed to continue unchecked, then the Earth's climatic zones are likely to shift. So, for example, temperate regions will become much warmer and the world's glaciers and ice masses will melt, leading to an overall rise in sea level.

The spread of arid zones may bring drought and water shortages to new areas.

Strategies suggested to reduce global warming include cutting back on the use of fossil fuels to reduce carbon dioxide emissions, the development of renewable, non-polluting energy sources, such as wind and solar power, and the halting of deforestation matched by a massive replanting program. The burning of felled trees adds carbon dioxide to the air, while forests remove carbon dioxide via the process of photosynthesis. Clearly, an increase or decrease in the extent of forests will influence the volume of carbon dioxide entering or being removed from the atmosphere.
See: Atmosphere, Carbon Dioxide, Chloroflurocarbons, Deforestation, Global Warming, Methane, Ozone Layer, Photosynthesis.

Gregale
The name applied to a strong wind which blows from the northeast in the central regions of the southern Mediterranean. It is most frequent in the cool season when there is a zone of high pressure over northern Europe and the Balkans, and an area of low pressure is in position over North Africa. The weather associated with a Gregale is little rain and a small drop in the average temperature.
See: Wind

Ground Frost
A term used in Great Britain to denote a surface temperature below freezing point and an immediate air temperature above freezing point.
See: Frost.

Growing Season
A term which describes the period of the year in which the vegetation specific to a particular climatic zone can flourish. Generally, the length of the growing season declines with distance from the equator and height above sea level. However, there are many exceptions to these general rules due to local climatic anomalies and because of humankind's ability to either deliberately or inadvertently modify the climate. Heated greenhouses are an obvious example of an artificial environment designed to increase the growing season.
See: Global Warming.

Gust
A shortlived and sudden increase in wind speed.
See: Beaufort Scale, Wind.

Haar
A type of sea fog found in the summer along the east coast of England. It is driven on shore by prevailing easterly winds.
See: Fog, Sea Fog.

Haboob
A duststorm found in the north and northeast of the Sudan in East Africa. It is most prevalent between May and September and is usually experienced in the afternoon and evening. The onset of an haboob is heralded by a sudden rise in wind speed and a change in its direction. The wind also brings a significant fall in temperature, a drop in visibility because of the dust carried by the haboob, and is usually followed by heavy rain or thunderstorms.
See: Duststorm, Thunderstorm, Wind.

Hadley Cells
Named after English scientist George Hadley (1685–1768), who announced his ideas in 1735, these are responsible for a planet-scale movement of wind in the middle and upper troposphere, both in the northern and southern hemispheres. Air descends in subtropical anticyclones and is borne toward the equator by surface trade winds. Here the air is heated and rises. Once it has risen to the middle and upper troposphere at the intertropical convergence zone, the air flows away from the doldrums poleward into the subtropical zones of high pressure. Because of the effect of Coriolis force, these upper winds veer toward the right in the northern hemisphere and to the left in the southern hemisphere. A similar process occurs at the poles where cold air at low level moves horizontally toward warmer zones. Here, the air is heated, rises, and returns to colder regions.
See: Anticyclone, Coriolis Force, Intertropical Convergence Zone, Doldrums, Troposphere, Wind.

Hail
Hard balls of ice with diameters above 5mm, they are produced in convective clouds of the cumulonimbus type and are formed by the rapid rise of a parcel of moist air. As the water drops in the parcel freeze, their size increases as more water vapor freezes on their surface. Once the individual ice pellets are heavy enough to overcome the effects of the rising air currents, they fall to the ground, and they may grow even larger as they fall by taking up even more of the supercooled water pellets they encounter during the descent. Hail is often associated with a thunderstorm.
See: Convection, Convection Cloud, Cumulonimbus, Thunderstorm, Water Vapor.

Hailstreak
An elongated pattern of hail on the ground.
See: Hail.

Harmattan
Found in West Africa, this is a strong northeasterly wind which blows from the Sahara desert. The harmattan is hot, dry, and dusty.
See: Wind.

Haze *(Below)*
A mass of small solid dust or smoke particles which reduces visibility below 1.25 miles, but not below 0.6 of a mile. Typically, haze has the appearance of a bluish tinge when viewed against a dark background and a yellow-brown color when seen against a lighter background. Hazes are common in calm conditions when the lack of wind prevents the dispersal of the particles and a temperature inversion, a zone in which temperature does not as normal decline with altitude but rather increases.
See: Fog, Mist.

Heat Equator

The latitude of the Earth's highest average annual temperature, the heat equator is located approximately 10 degrees north of the geographical equator. The northern hemisphere is marginally warmer on average than the southern hemisphere for several reasons. Because of the greater extent of the ice cover of the Antarctic, more solar radiation is reflected back as the ice has a greater reflective value, thereby lowering the temperature. Greater sea cover in the Arctic generates higher temperatures as the water is warmed up and stores heat which is given off slowly. Secondly, the larger area of tropical land in the northern hemisphere warms up more quickly. Finally, prevailing ocean currents favor the movement of warm water into the northern hemisphere.
See: Equator.

Heatwave

A term used to refer to a prolonged period of unusually hot weather well above the average experienced in a particular region.
See: Temperature, Weather.

Helium

A colorless, light, nonflammable inert gas found in the atmosphere and comprising 0.00052 percent of the total.
See: Atmosphere.

Helm Wind

This is a cold, strong wind which is prevalent in the north of England during later winter and spring and is particularly common when the wind is easterly or northeasterly. Its name is taken from the dense, helmet-shaped cloud which is associated with the wind and is found over high ground.
See: Banner Cloud, Cloud, Wind.

High

A zone or region of high atmospheric pressure such as an anticyclone.
See: Air Pressure, Anticyclone, Atmosphere.

Hill Fog

A term used by meteorologists to describe low-lying cloud, that has the appearance of fog, found covering high ground.
See: Cloud, Cloud Classification, Fog.

Hoar Frost *(Below)*

Consisting of feathery and white interlocking ice-crystals deposited on surfaces that have been cooled below freezing, usually by heat loss through radiation, hoar frost has two components: dew which has been deposited on a surface and then frozen and water vapor from the atmosphere that has frozen directly onto a surface. As smaller objects, such as leaves, tend to cool much more speedily that larger ones, hoar frost tends to form most quickly on the edges of these.
See: Deposition, Frost, Water Vapor.

Horse Latitudes

These are zones of calms and light winds associated with dry conditions and stable weather. They are found in the belts of high pressure in the subtropical zones over the oceans in both the northern and southern hemispheres, situated between the trade winds and westerlies, and shift north or south as the Earth's position relative to the sun changes. The name is allegedly derived from the long-gone practice of throwing horses overboard from sailing ships becalmed in the region so as to lighten the load and make the best of the light winds.
See: Trade Wind, Westerlies, Wind.

Humidity

Humidity is a catch-all term which refers to a variety of measures by which the atmosphere's water vapor can be expressed.

See: Atmosphere, Absolute Humidity, Relative Humidity.

Hurricane *(Right, pages 92/3 and 94/5)*

Also known as a typhoons, hurricanes are cyclones that develop over the oceans in the tropics and move inland often causing extensive damage and loss of life. The names is taken from Huracan, a god of evil. Hurricanes develop over the oceans north and south of the equator, beginning as a group of extremely violent thunderstorms. Energy is drawn from the warm oceanic waters and, under the right conditions, an intense system of enclosed air movement around a zone of low pressure develops. The system then develops further, with the pressure in the center falling even more, the winds gathering speed, and the hurricane beginning to head away from the equator. If the wind associated with the weather system reaches 23mph, it is termed a tropical depression. When the winds reach a speed of 39mph, the system is called a tropical storm and given a codename; if the speed then rises to more than 73mph, it gains official recognition as a hurricane. Hurricanes tend to move westward at a speed of 6–12mph, but the direction of their movement can be erratic. Hurricanes are not permanent weather systems and tend to decay and then disappear after eight days or so, but the winds and heavy rains linked to them are frequently highly dangerous. A scale of intensity and the possible damage associated with hurricanes has been developed by H.S. Saffir and R.H. Simpson. This, the Saffir-Simpson Hurricane Intensity Scale, grades hurricanes from one to five, with the latter figure representing the most destructive type. Each of the five degrees of intensity identify a hurricane on the basis of four factors: the range of its central air pressure, the range of its wind speed, the height of the associated coastal storm surge, and the likelihood of property damage. In the latter case, this ranges from minimal for a scale one hurricane to catastrophic for a scale five hurricane. Hurricane Andrew in 1992 was given a scale number of four, indicating that areas in its path could expect extreme damage. Hazards linked to hurricanes consist of high winds, heavy rainfall and subsequent flooding, and storm surges which can flood

coastal areas. Hurricane Andrew, which came ashore at Miami Beach, Florida on August 24, 1992, brought winds that gusted to over 165mph and caused massive damage to property. Over 180,000 people were made homeless and 25 died, as the hurricane cut a swath of destruction across southern Florida. The damage to property in Florida alone was estimated at more than $25 billion. Trees were uprooted, moored boats were tossed ashore and smashed like so much matchwood, buildings had their roofs torn off or they were obliterated, and trailers were thrown in the air and dashed to the ground. Andrew, however, was far from finished and continued westward across the Gulf of Mexico, striking the coast of Louisiana two days later. Again the damage was severe, estimated at $400 million dollars, and four residents lost their lives.

See: Cyclone, Eye of the Hurricane, Tropical Cyclone, Typhoon

Hydrological Cycle *(Illustrated On page 98)*

Sometimes referred to as the water cycle, this is the movement and exchange of water between the Earth's land, oceans, and atmosphere. The total volume of water contained within these components has varied little throughout the Earth's history, although the volume of water moving between the three does vary. The most obvious case of water taking part or being removed from the cycle is during periods of glacial advance or retreat. Ice sheets and glaciers are water stores and reduce the

volume of water able to switch between land, sea, and air. The current distribution of water is as follows:

Water Reservoir	Percentage Held
Oceans	97.2
Glaciers and ice sheets	2.15
Groundwater	0.62
Freshwater	0.009
Inland seas and salt lakes	0.008
Soil water	0.005
Atmosphere	0.001
Rivers and streams	0.0001

Water, in various forms, is constantly switching between these reservoirs. The key movements are water being turned into vapor through evaporation from the land and sea which then enters the atmosphere to form clouds, which return water to the land and sea through precipitation. Water is exchanged between land and sea via rivers and streams or by direct runoff.
See: Evaporation, Evapotranspiration, Precipitation.

Hydrosphere

This term refers to the total amount of water in both solid and liquid form on the Earth's surface and the volume of water to be found in the atmosphere.
See: Atmosphere, Hydrological Cycle.

Hydrostatic Equilibrium

This is the balance between the atmosphere's vertical pressure gradient (pressure declines at a constant rate with increasing altitude) and gravity. If these two forces are equal, an air mass does not undergo a change in its velocity of movement. However, this does not prohibit vertical movement, either upward or downward, because a rising parcel of air will continue to rise at a constant velocity, as would a descending parcel of air.
See: Pressure Gradient, Gravity.

Hyetograph

A type of self-recording rain gauge consisting of a container fitted with a float which rises as the vessel fills with precipitation. As the float rises, a tracing device attached to it records the rainfall total on a chart fixed to a rotating drum.
See: Precipitation, Rain Gauge.

Hygrogram

Recorded on a hygrograph, this is the continuous record of changes in the atmosphere's relative humidity, usually over a period of a week.
See: Atmosphere, Hygrograph, Relative Humidity.

Hygrograph

A form of hygrometer in which the relative humidity of the atmosphere is continuously recorded on a chart fixed to a clockwork rotating drum. The somewhat outdated form of this instrument consists of a human hair which expands or contracts as the relative humidity changes, and the variations are then recorded via a tracing pen on to the rotating chart, known as a hygrogram. This is a somewhat crude device, but is of some value in recording the time and scale of sudden or large changes in relative humidity.
See: Atmosphere, Hygrogram, Relative Humidity.

Hygrometer

A device used to measure the atmosphere's relative humidity consisting of a dry-bulb thermometer, which measures the actual temperature of the atmosphere, and a wet-bulb thermometer. The bulb of the latter is kept permanently wet and evaporation of moisture from the thermometer leads its temperature always to be lower than that of the dry-bulb thermometer. The difference between the two readings gives a crude measure of the air's relative humidity.
See: Atmosphere, Hygrograph, Relative Humidity, Thermometer.

Hygroscope

A device used to show variations in the humidity of the air, most frequently through changes in the appearance or dimensions of a substance which is sensitive to changes in humidity.
See: Humidity.

Hythergraph

A means of indicating the differences between different types of climate plotted on a graph. A hythergraph consists of average monthly temperatures which are plotted on the vertical scale and the average monthly rainfall or humidity as the horizontal coordinate. The resulting 12-point closed polygon is the hythergraph, a device from which a climate type can be identified at a glance.
See: Climate, Climatic Zones.

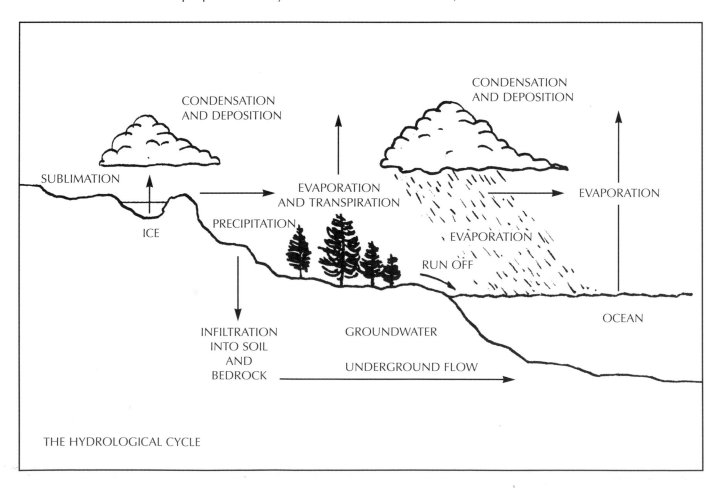

THE HYDROLOGICAL CYCLE

Ice *(Below)*

The solid form of water when its temperature falls below 0°C (32°F). It has a density of about 0.9 grams per cubic centimeter. Approximately 99 percent of the world's ice is found in the ice sheets of Antarctica and Greenland, and other sources include mountain glaciers. In terms of the weather, two types of precipitation, snow and hail, are also composed of ice. Frost also consists of ice crystals.
See: Frost, Hail, Ice Age, Ice Cap, Precipitation, Snow.

Ice Age

Although comprehensive data on the Earth's climate does not exist for more than the last 100 years, scientists have been able to reconstruct partially climates of the very distant past. Evidence suggests that the Earth has undergone a number of ice ages which were separated by a number of warmer, interglacial periods. Although there is evidence of ice ages as far back as more than 250 million years before the present, information on more recent ice ages is much stronger. The most recent glacial period began some 25,000 years ago, reached its maximum extent some 18,000 years ago, and retreated some 10,000 years ago. It is suspected that changes in the energy output of the sun or variations in the Earth's orbit around it probably played a part in the growth and decay of the world's ice sheets.
See: Climatic Change, Greenhouse Effect, Global Warming.

Ice Cap

A general term referring to the ice sheets that cover the North and South Poles.
See: Ice Sheet.

Ice Cap Climate *(Illustrated pages 100/1)*

One of the Earth's climatic zones associated with the massive ice caps and sheets of the northern and southern hemispheres, chiefly Antarctica and Greenland. Inhospitable to humankind, the climate is dominated by low temperatures (the mean monthly temperature is below freezing point), there is little precipitation and what does occur falls as snow, and often high-speed winds bring blizzards. The extent of the ice caps have varied through time due to natural and, much more

recently, manmade climatic change. As the Earth's climate cools, the glaciers advance, lowering the sea level; as heating occurs the ice caps retreat, raising the sea level.

See: Climatic Zones, Climatic Change, Global Warming.

Ice-Forming Nuclei *(Left)*

These are minute particles, such as those generated by burning fossil fuels, in the atmosphere which aid the creation of ice crystals at temperatures considerably below freezing point. There are two distinct types of ice-forming nuclei: water vapor is deposited as ice on deposition nuclei, while freezing nuclei encourage water droplets to freeze.

See: Cloud Condensation Nuclei, Deposition, Water Vapor.

Ice Pellets

Water in the atmosphere exists in three forms: liquid, solid, and gas. In its solid form, water consists of ice crystals of various sizes and varying degrees of hardness. Ice pellets comprise frozen rain droplets with a diameter of below 5mm and are more commonly known as sleet. Sleet forms when rain descends from a comparatively mild layer of air and falls through a deep layer of extremely cold air which freezes the water before it strikes the ground.

See: Hail, Precipitation.

Ice Sheet

A very extensive expanse of flat, compacted ice and snow covering land masses in the Earth's higher latitudes, chiefly Antarctica and Greenland. The thickness of the ice varies but can reach several thousand feet.

See: Ice, Ice Cap, Ice Cap Climate.

Impaction

This is a natural process that removes aerosols from the atmosphere through the mechanism of them striking and adhering to structures such as buildings.

See: Aerosols, Gravitational Settling.

Indian Summer

An Indian Summer is a type of weather singularity, meaning that it is a weather episode that occurs

regularly at roughly the same time on each occasion and is out of keeping with the supposed normal weather to be expected at that time. However, there is no precise date for the beginning or end of the Indian Summer phenomenon, which takes place in North America and Great Britain during fall/autumn. The weather associated with this singularity is mild with sunny days and cool, frosty evening and nights. In North America, it is associated with large anticyclones that fix over the eastern margins of the continent and block the weather systems that are associated with more usual seasonal weather.

See: Anticyclone, Autumn, Fall.

Insolation

This is the name of the solar radiation reaching the Earth and is a contraction of incoming solar radiation. Of the radiation that reaches the planet from the sun, some 45 percent is visible as sunlight, while the rest is invisible and consists of 46 percent infrared radiation and nine percent ultraviolet radiation. The amount of insolation striking a particular point on the Earth's surface is a function of the solar constant, the inclination of the point to the sun's rays, and the degree of transparency of the atmosphere, and the orbit of the Earth around the sun. Generally, the intensity of radiation striking the Earth is greatest at the equator and is at its lowest at the North and South Poles.

See: Equinox, Solstice, Radiation, Solar Constant.

Instability

Warm air rises, expanding and cooling as it ascends through the atmosphere, but how far it rises and where it halts is dependent on the properties of the air surrounding the parcel of warm air. In the case of a parcel of rising warm air being surrounded by a mass of considerably colder air, the warm air will continue to rise, despite its own expansion and cooling because the surrounding air is still considerably cooler. The rising warm air parcel may even accelerate its rate of ascent if it reaches its dew point and condensation is initiated. Condensation releases latent heat, therefore

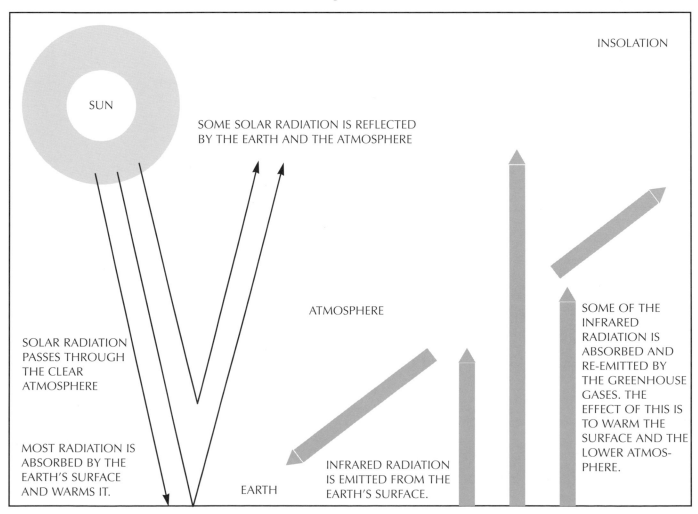

INSOLATION

SUN

SOME SOLAR RADIATION IS REFLECTED BY THE EARTH AND THE ATMOSPHERE

ATMOSPHERE

SOLAR RADIATION PASSES THROUGH THE CLEAR ATMOSPHERE

MOST RADIATION IS ABSORBED BY THE EARTH'S SURFACE AND WARMS IT.

EARTH

INFRARED RADIATION IS EMITTED FROM THE EARTH'S SURFACE.

SOME OF THE INFRARED RADIATION IS ABSORBED AND RE-EMITTED BY THE GREENHOUSE GASES. THE EFFECT OF THIS IS TO WARM THE SURFACE AND THE LOWER ATMOSPHERE.

increasing the temperature contrast between the warm, rising air and the surrounding cool air. This process is referred to as instability and is associated with the formation of cumuliform clouds.
See: Condensation, Cumuliform Cloud, Dew Point, Stability.

Insular Climate

A type of prevailing climate that is found on islands or in coastal regions in which the impact of oceans is greater than a large land mass. Although insular climates vary widely, they exhibit a generally small daily and annual temperature range.
See: Climatic Zones, Continental Climate, Maritime Climate.

Intertropical Convergence Zone

This is a fractured belt of thunderstorms which runs parallel to the equator in both the northern and southern hemisphere and marks the point on the Earth's surface where trade winds converge. It is an area of generally light winds, known as the doldrums. Large-scale planetary winds meet at the convergence zone, which is usually to be found at the heat equator, and rise into the upper atmosphere. They are accompanied by an increase in convective cloud cover and rain. The ITCZ is an important component of the Hadley Cells which circulate air masses between the poles and the equator in both hemispheres.
See: Convection, Convergence, Doldrums, Hadley Cells.

Inversion of Temperature

Under normal circumstances temperature declines with height above the ground. However, a temperature inversion is the reverse of this general trend; that is temperature rises with altitude. Inversions are not confirmed to any particular zone within the Earth's atmosphere and can be found at low and middle altitudes. At the surface, temperature inversions can develop in depressions, hollows, and valleys in which radiation of heat on cold and clear nights leads to parcels of cold air sinking into the low-lying ground, while parcels of warmer air are positioned on the slopes above. Consequently, the temperature is lower at low levels and warmer at higher levels.
See: Haze.

Ionosphere *(Below and page 108)*

This refers to the zone in the Earth's upper atmosphere which ranges from 50 to 140 miles above the surface and contains high concentrations of ions, electronically charged particles. The ionosphere has little influence on the weather, but it is the site for the northern lights (aurora borealis) and the southern lights (aurora australis). These are seen in the night sky and consist of blankets of green-white light.
See: Atmosphere.

Isallobar

Generally found on weather charts, isallobar are lines connecting points which have undergone the same changes in atmospheric pressure over the same period of time. The individual readings are calculated by working out the change of pressure that has taken place between two set points in time. The resulting charts are useful devices for identifying areas of rising or falling pressure.
See: Air Pressure, Atmosphere, Barometer.

Isanomalous Line

This a line on a chart which joins points that have had an equal departure from the norm of a partic-

ular meteorological measure, for example, temperature or rainfall. The impact of latitude and altitude are removed from the readings. Areas that have an above average degree of cold can be described as having a negative temperature anomaly, while those having a below average degree of cold can be said to have a positive temperature anomaly.
See: Temperature.

Isobar

A line on a chart linking points that have equal atmospheric pressure. It is usual for the readings gathered at a weather station to be amended to sea level so that the influence of altitude on pressure is removed. On weather maps, isobars are drawn to reveal the distribution and variations in atmospheric pressure at a chosen time on a particular day in a reasonably small area, such as a country or continent. Generally, the isobars are drawn at two or four millibars on a daily weather map and usually even numbered readings, for example 998mb, 1000mb or 1004mb and 1008mb, are shown. When climatologist produce maps of atmospheric pressure, they are looking at long-term trends and tend to drawn isobars on world maps, although this is not always the case. Isobar derives from two Greek words *iso* meaning equal and *baros* meaning weight.
See: Air Pressure, Atmosphere, Climatology, Millibar.

Isohel

A line on a chart that connects points having equal amounts of sunshine over a certain period. *Helios* is the Greek for sun.
See: Sun, Sunshine.

Isohyet

A line on a chart that connects points that have had an equal amount of rainfall over a certain period. Hyet, meaning "indicating rain," is taken from the Greek for rain, *huetos.*
See: Precipitation.

Isoneph

A line on a map that links places that have an equal mean of cloudiness over a certain period. Neph, meaning "concerning clouds," comes from the Greek for cloud, *nephos.*
See: Cloud, Cloudiness.

Isotherm

This is a line on a weather chart that joins points having the same temperature at a particular moment of time or, more usually, having equal mean temperatures over a given period. Readings are normally reduced to sea level to negate the effect of altitude. There are two types of isotherm that are of particular value to a meteorologist. The first links points that have a similar average temperature over a period of six months. The readings are normally taken in January and July, and therefore give average temperature conditions in the middle of winter and summer. The second isotherm map type records either the mean daily maximum or minimum temperatures.
See: Temperature.

Jet Stream

Located at the level of the tropopause, the boundary in the Earth's atmosphere found between the troposphere and the stratosphere, two of these strong, high altitude winds have been identified flowing from west to east in narrow corridors: the polar front jet stream of the middle latitudes and the subtropical jet stream. These powerful winds are generated by the strong pressure gradients which reflect the great temperature differences at high altitudes. The importance of the jet streams lies in their key role in the development and movement of low pressure systems. The jet streams vary in intensity, tending to be at their strongest during winter, and they often fragment or move north or south. The polar front jet stream follows a winding path and can reach speeds of well over 100mph. This jet stream is found along the polar front, a narrow region situated between cold and warm air masses. The subtropical jet stream is located between the midlatitudes and tropics and is more fixed than its northern counterpart and has stronger winds.
See: Air Mass, Atmosphere, Hadley Cell, Stratosphere, Tropopause, Troposphere, Wind.

Karabuan

Produced by the rapid heating of the interior of the Central Asian land mass, this hot and strong north-easterly wind blows across central China's Sinkiang Province from early spring until the end of summer, chiefly in the daytime. The wind, often gale force, gathers up large clouds of fine dust

from the desert floor, which darken the sky and carry the particles over long distances before they are deposited.
See: Continental Climate, Duststorm, Wind.

Katabatic Wind

A localized wind produced at night by the flow of air cooled by radiation down mountains or along valleys, or by the flow of cold air over the surface of ice masses, such as those of Antarctica and Greenland. Because of the rapid loss of heat by radiation, the mountain, valley, or ice mass cools rapidly, cooling the air above in turn, which then moves downhill under the effect of gravity. The direction of flow is governed by local topography.
See: Anabatic Wind, Mountain Wind, Radiation, Wind.

Khamsin

The Arab word for 50, khamsin is also the name given to a hot, strong, and dry southerly wind that is found in Egypt and is traditionally believed to blow for 50 days each year between April and June. The khamsin, which carries large quantities of dust and sand particles from the North African interior, is found ahead of the depressions which track eastward across the Mediterranean Sea or follow the North African coastal fringe. The name is now widely used in the Middle East to describe any similar wind blowing from a desert area.
See: Depression, Wind.

Knot

This unit of speed, often used to measure wind velocity, is equal to one nautical mile (1.15 miles) per hour.

Koembang

Found in Java in the Southwest Pacific, this wind is warm and dry.
See: Fohn, Wind.

Krypton

An inert gaseous element found in the Earth's atmosphere, accounting for 0.00010 percent of the total by volume.
See: Atmosphere.

Lake Breeze

A type of low-speed wind which blows from an inland water surface on to land and is produced by the different rates at which land and sea warm up and cool down throughout the day. The land generally warms up more quickly than water and, because of this, a localized area of lower pressure develops over the warmer surface of the land, while an area of relatively higher pressure exists over the cooler water. Thus there is a horizontal pressure gradient running from the water to the shore and winds blow down these, bringing cool, often refreshing breezes to the shoreline and the adjacent inland area. Lake breezes tend to develop during the daytime and decay during the evening as the land cools more rapidly than the water.
See: Breeze, Land Breeze, Pressure Gradient, Sea Breeze.

Lake-Effect Snow

Most frequently found in winter and autumn/fall, lake-effect snow consists of a local fall of snow downwind of an area of open water. Such snow is a product of localized air mass modification. This involves cold air passing over a warmer lake, thereby encouraging evaporation and reducing the stability of the cold air. Convection therefore takes place and clouds develop. The now-modified cold air blows onshore, where it slows and begins to rise above the cold land air. Cloud development increases and snow falls. The snowfall totals vary with the nature of the lakeside topography. If hilly, the air is forced higher, cloud development is greater and the snowfall becomes heavier. The snowfall associated with this feature is often of short duration but can be extremely intense. These episodes, known as snowbursts, can bring chaos to roads and disrupt the movement of traffic.
See: Air Mass Modification, Convection, Evaporation.

Land Breeze

This breeze is similar to a lake or sea breeze and is produced by the differential heating and cooling properties of land and water. In the daytime, land heats up more quickly that water, but it cools more rapidly during the evening and night. At night, the radiation cools the air above the land, while the air over the water surface remains warmer. Consequently, a pressure gradient develops in which an area of high pressure is created over the land and an area of relatively low pressure is established over the water. Winds blow from high to low pressure and so a breeze flows from the land to the water. Land breezes tend to be at their strongest just before sunrise when the pressure gradient is at its greatest.
See: Breeze, Lake Breeze, Pressure Gradient, Sea Breeze.

La Nina

Found in the central and eastern tropical zones of the Pacific, La Nina consists of a period of strong trade winds and low sea temperatures and occurs about every 3–7 years. The phenomenon is a product of a change in the pressure gradient in the region which allows for the development of colder than normal surface waters in the Pacific eastern tropics and warmer ones in the western tropics. Its causes are the opposite to those that produce the anomaly known as El Nino.
See: El Nino.

Latent Heat

This term relates to the movement of heat from one point to another due to a change in water from solid to liquid to gas, or vice versa. Whether water gains heat from the environment or transfers heat it depends on the nature of the change occurring to it. Latent heat is released into the environment when water vapor undergoes deposition, or condenses to form a liquid, or freezes to produce ice. Latent heat is gathered from the environment when ice melts to form water, undergoes sublimation to produce water vapor, or when water evaporates to form water vapor.
See: Condensation, Deposition, Sublimation, Water Vapor.

Lee

The side of an object, for example, a mountain range, which is sheltered from the prevalent direction of the wind.
See: Fohn, Wind

Lenticular Cloud

A cloud with the appearance of a wave or lens. Lenticular clouds have well-defined edges, are

most commonly visible in upland or mountainous areas, and are often associated with the fohn wind. These cloud types most generally develop when the wind remains constant both in speed and direction The cloud formations most often associated with a lenticular shape are altocumulus, cirrocumulus, and stratocumulus
See: Altocumulus, Cirrocumulus, Cloud, Cloud Classification, Fohn, Stratocumulus.

Leste

Commonly felt by the inhabitants of the island of Madeira lying off the northwest coast of Africa, this hot, dry, and dust-carrying wind occurs in advance of a passing depression. It is a southerly to easterly wind.
See: Depression, Wind.

Levanter

A wind, sometimes called the solano, blowing across the Straits of Gibraltar at the western entrance to the Mediterranean Sea and also found in southern Spain. It is particularly common in late summer. The levanter gained its name because of the direction from which it blows; the Levant was the former name of a portion of the eastern Mediterranean which includes modern Lebanon, Syria, and Israel.
See: Wind.

Leveche

A hot, southerly, dust-laden wind experienced in southeast Spain which is found in the van of an advancing depression. The sirocco of North Africa is a similar phenomenon.
See: Depression, Sirocco, Wind.

Lightning *(Left and pages 112/3, 114)*

Lightning occurs when the violent ascending air currents found in certain cloud types, particularly well-developed cumulonimbus and, to a lesser extent, altocumulus, act to separate positive and negative charges within the cloud. The positive charges tend to move toward the upper reaches of the cloud while the negative charges are forced into its lower reaches. Under these circumstances, the cloud also induces positive charges on the ground immediately beneath it. This moves with the cloud. Lightning occurs when the charges become so strong that they can overcome the nat-

ural resistance of the air and discharge between positive and negative charges takes place. This is the lightning of which two main types have been identified: fork lightning and sheet lightning. Fork lighting, which has the appearance of tree roots or branches, takes place between the negatively charged base of the cloud and the positively charged ground. Sheet lighting occurs between the positive and negative charges within a cloud and gains its name because the lightning strike is masked by the cloud itself and the flash illuminates a large section of the cloud.

Lightning is accompanied by thunder and it is possible to estimate a viewer's distance from a thunderstorm by noting the time lapse between the flash of the lightning and the arrival of the sound of the thunder. Light travels faster than sound so, for example, a gap of five seconds between the flash and the sound indicates that the thunderstorm is about one mile away. There are a number of other types of lightning, including rocket lightning, which shoots from the upper reaches of a cloud into the atmosphere above.

See: Altocumulus, Cumulonimbus, Thunderstorm.

Line Squall

The location of a line squall is associated with the replacement of a current of warm air by a much cooler air mass and the arrival of a cold front of a depression. As the warm air is forced aloft by the denser cool air, condensation and cloud development are initiated. A line of squalls, violent winds that grow and die away rapidly, develops along this front. Squall lines can stretch for many miles and are often accompanied by dark clouds at low altitudes which produce heavy rain and hail.

See: Cold Front, Front, Depression Hail, Squall.

Lowering

This is one of the first features associated with the development of a tornado. A lowering comprises a dense mass of cloud which grows beneath the main cloud mass. The lowering cloud has distinctive rotational and vertical movement because air is being drawn up into the tornado's main cloud mass.

See: Cloud, Funnel Cloud, Tornado.

Low

A term referring to an area where the atmospheric pressure is noticeably lower that its surroundings or a depression or cyclonic weather system. On weather charts such features are marked by use of the world low. The opposite of a low is a high, which is linked to a zone of high atmospheric pressure and weather systems such as an anticyclone.

See: Anticyclone, Cyclone, Depression, High.

Lull

A brief drop in the speed of a wind.

See: Wind

Mackerel Sky *(Illustrated pages 116/7)*

A phenomenon usually associated with the dry and warm weather of summer, a mackerel sky consists of a mass of altocumulus or cirrocumulus clouds which is fractured into a pattern of ripples which are supposedly suggestive of the banded markings found on the side of mackerel.
See: Altocumulus, Cirrocumulus, Cloud, Cloud Classification.

Macroburst

A means of classifying a downburst, the strong downdraft of wind occasionally associated with a thunderstorm and responsible for destruction of property. A macroburst downdraft wind travelling horizontally once it has hit the Earth's surface at speeds of over 130mph causes destruction over a distance of more than 2.5 miles and can last up to 30 minutes.
See: Downburst, Microburst, Thunderstorm.

Maestrale

Found in northern Italy, this is a cold northerly to northwesterly wind similar to the mistral which is found in the northwestern Mediterranean.
See: Mistral, Wind.

Maize Rains

The name given to the heavy and prolonged rains that fall across East Africa between February and May.
See: Millet Rains, Precipitation.

Mammatus Cloud

Most commonly applied to cumulonimbus clouds, mammatus types can be identified by their hanging pockets and usually develop at the lower levels of a thunderstorm. Mammatus clouds develop when pockets of cold and moist air begin to descend from a thunderstorm and come into contact with unsaturated air. Through the process of sublimation, ice crystals at the margins of the descending pockets of cold air vaporize, thereby cooling the air further and offsetting the effects of compressional warming. This cooling, cloudy air continues to fall through the warmer, lower air producing the associated hanging protuberances.
See: Cloud, Cloud Classification, Compressional Warming, Sublimation.

Mango Showers

Also known as blossom showers, these are the rains that fall on the coffee cultivating regions of Southeast Asia between March and May.
See: Blossom Showers, Precipitation.

Mares Tails

A common term used to identify wispy cirrus clouds.
See: Cloud, Cloud Classification.

Maritime Climate

A climate dominated by the effects of seas and oceans, and typically consisting of a cool, mild summer and a warm winter. Variations in seasonal and daily temperatures are generally less than those found in the interiors of continents. Because of its thermal properties, a large water mass tends to heat up more slowly during the day and over summer and cools more slowly at night and over winter than does a large land mass. Consequently, land areas lying downwind of an expanse of water, have climates dominated by the air masses that take on the properties of the waters over which they form or pass. Western Europe has a typical maritime climate because west-to-east winds blow maritime air masses over the continent.
See: Air Mass, Air Mass Modification, Continental Climate.

Maritime Polar Air

A type of air mass that forms over the cold waters of the North Atlantic and North Pacific. Over the North Pacific, maritime polar air is generally mild and humid in winter and summer, while over the North Atlantic, it is cold and humid in winter and cool and humid over the summer. On weather charts a mass of maritime polar air is identified by the letters mP.
See: Air Mass, Air Mass Modification.

Maritime Tropical Air

A type of air mass that forms over the world's oceans in the tropics and subtropics. These air masses, found in both the Atlantic and Pacific, are warm and humid in both winter and summer.
See: Air Mass, Air Mass Modification.

Maunder Minimum

Named after E. Walter Maunder, this refers to the period between 1645 and 1715 when the sunspot activity of the sun was greatly reduced and coincided with noticeably colder climates in Western Europe. Little notice was taken of Maunder's view that increases and decreases in sunspot activity could affect the Earth's climate until the 1970s, when the relationship came under greater scrutiny. The general principle is than an increase in sunspot activity leads to a brighter sun which gives of more radiation which in turn warms the Earth's climate. Correspondingly, a fall in sunspot activity leads to a cooler sun, less radiation, and a cooling of the Earth's climate. However, the exact relationship between sunspots and the Earth climate is still far from clear. Evidence that might undermine the supposed relationship include findings that the cold spell was not global and that other cold spells have not coincided with reduced sunspot activity.

See: Climatic Change, Sporer Minimum, Sunspot.

Mediterranean Climate

Originally used to identified the climates found in the land areas bordering the Mediterranean Sea, this term is also used to define climates in many other areas, including a narrow coastal strip of central California, the southern fringe of South Africa, and parts of southwest and south Australia. Generally, Mediterranean climates are found on the western edges of large continental land masses between 30 and 45 degrees latitude both north and south of the equator and are typified by warm, dry summers, and mild, wet winters. Annual rainfall totals range from 16 to 31 inches. Average yearly temperatures vary with distance from the coast: higher summer temperatures are found inland, where the cooling effects of inshore sea breezes are minimal, and, because of this, there is a more pronounced variation in temperatures between summer and winter than in coastal areas.

See: Climatic Zones

Meltemi

The name given in Turkey to the strong northerly Etesian winds found in the eastern Mediterranean during summer.

See: Etesian Winds, Wind.

Mesocyclone

A transitional stage, part of the process in which a thunderstorm develops into a tornado. In a violent thunderstorm the horizontal wind speed increases

and the wind turns clockwise as height above the ground rises. This causes the wind to begin to rotate about the horizontal axis and, when this motion reacts with the updraft of air, it is tilted into the vertical. This adds to the rotation of the air about the vertical and the updraft rotates anticlockwise. This forms the mesocyclone from which tornadoes can develop.
See: Thunderstorm, Tornado.

Mesosphere

One of the vertical divisions of the Earth's atmosphere situated between the stratopause and the mesopause. It extends from approximately 30 miles above the surface to approximately 50 miles, and is a zone in which temperature falls with altitude from approximately freezing point at the lower levels to minus 80°C in the upper reaches.
See: Atmosphere

Meteorology

This is the study of the Earth's atmosphere and the phenomena that generate the weather, which is defined as the particulars of the atmosphere, such as cloud cover, precipitation, temperature, and wind speed, at a particular time at a particular place.
See: Climate, Climatology, Weather.

Methane

A colorless, odorless, and flammable gas in the atmosphere that makes up approximately 0.00002 percent of the total by volume. Although methane occurs naturally, its concentrations in the atmosphere are rising due to humankind's activities, especially rice production, animal husbandry, and waste disposal. It is believed that methane is a key component in global warming. Estimates in the 1980s suggest that methane was contributing 18 percent to the greenhouse effect, second only to carbon dioxide (50 percent).
See: Atmosphere, Carbon Dioxide, Global Warming, Greenhouse Effect.

Microburst

A means of classifying the destruction wrought by the downburst and violent horizontal winds associated with a thunderstorm. A microburst is of generally short duration, up to 10 minutes, and covers an area of less than 2.5 miles. Speeds can be over 150mph and cause extensive damage to property.
See: Downburst, Macroburst, Thunderstorm.

Microclimate

A climate usually restricted to a small area, such as a valley.
See: Climate.

Midlatitude Westerlies

These are planetary-scale winds that blow from west to east in middle to upper zones of the troposphere between 30 and 60 degrees latitude both north and south of the equator.
See: Troposhere, Wind.

Milankovitch Cycle

First proposed by Serbian scientist Milutin Milankovitch in the 1920s and 1930s, this cycle related to the geometrical relationship between the Earth and the sun and has been used to explain variations in the Earth's climate as well as variations in the amount of insolation striking the planet's surface both seasonally and with changes in latitude. Milankovitch identified three factors that could produce cyclical variations in the Earth's climate: first, changes in the tilt of the planet's axis. As the tilt increases, the contrast between summer and winter becomes greater. Secondly, he identified changes in the path followed by the Earth around the sun. The more that this path departed from a perfect circle, then the amount of insolation striking a point on the Earth would vary as it traveled around the Sun. Third, the wobble of the Earth's axis of rotation has varied through time, thereby affecting the times when a point on the Earth is either nearest to or farthest away from the sun. The effect of this would be to decrease seasonal climatic changes in one hemisphere, while increasing them in the other.
See: Climate, Insolation.

Millet Rains

The name given to the period of heavy rainfall found in East Africa between October and December.
See: Maize Rains, Precipitation.

Millibar

This is a unit of pressure that is the equivalent to one-thousandth of a bar, the unit of atmospheric pressure equal to 29.53mm of mercury at 0°C at 45 degrees latitude. On weather charts, isobars, the lines connecting points of equal pressure, are usually drawn at intervals of two or four millibars.
See: Isobar.

Mirage

This is an optical illusion in which images of distant objects can be seen quite clearly, mirror images can be produced, some objects can be elongated, or can be so distorted as to be barely recognizable. Mirages are created when bands of air with steep temperature differences lie above either a very hot or a very cold surface. Two types of mirage have been identified on the basis of the position of the object in the mirage in relation to its actual position. In inferior mirages, the object appears lower in the mirage than it really is; in superior images, the object appears higher in the mirage than it actual is.
See: Fata Morgana, Inversion of Temperature.

Mist *(Right)*

Mist is a type of low-lying cloud that, by generally accepted convention, restricts vision at distances beyond 3,250ft. Below this distance, the cloud is defined as fog. Both fog and mist develop when the temperature of parcel of air at ground level reaches dew point and condensation begins. There are a number of processes by which an air parcel can become saturated, including advectional, expansional, or radiational cooling. As temperatures fall, the first water droplets to condense are small and well spaced (mist), but, if the cooling continues, they become more numerous and more closely spaced (fog). The process is reversed as temperatures rise.
See: Advection, Condensation, Expansional Cooling, Fog.

Mistral

This cold northerly or northwesterly katabatic wind is found along the coast of the northwestern Mediterranean, particularly down the valley of the River Rhône in southern France. The wind is com-

mon during winter, when there is an area of high pressure associated with snow-capped mountains in position over the interior of Europe and an area of low over the western Mediterranean. In the case of the Rhône valley, air blows southward from the country's central plateau, runs down the narrow valley picking up speed, and hits the coast at the Gulf of Lyons as a powerful cold, dry wind. Speeds vary but can reach over 80mph. Skies associated with the mistral are often lacking cloud and temperatures often drop below freezing.
See: Bora, Katabatic Wind.

Monsoon

Deriving from the Arab word for season, *mawsim*, monsoons are climatic episodes found predominantly in Africa and Asia and are characterized by marked seasonal reverses in the direction of prevailing winds. These seasonal changes bring very wet summers and comparatively dry winters. In the case of India, some 75 percent of the region's annual rainfall arrives during the monsoon season which lasts from June to September.

It is believed that the onset of the monsoon season is related to seasonal movements in the circulation of planet-scale winds, particularly north-south movements in the intertropical convergence zone, and seasonal contrasts in the warmth of land and oceans. Evidence suggests that in spring the contrast between the relatively cool air over oceans and the warmer air over land leads to the development of a pressure gradient which allows humid air from the oceans to flow inland. Once on land, the air parcel is heated, begins to rise by convection, and becomes humid. As it rises, expansional cooling promotes condensation and the generation of clouds and precipita-

tion. Latent heat is released during this phase and the air continues to rise, thereby promoting even greater precipitation. At high altitude, the parcel of air heads back toward the sea, where it descends due to the water's cool surface. In autumn, the pressure gradient is reversed with the prevailing wind direction from land to ocean. Air descends over the land and then heads for the oceans, where it is warmed and rises.

The timing and intensity of the monsoon is influenced by physical geography. In the case of India, the wet monsoon period is influenced by the mountain ranges to the north of the country. During the winter, the high-altitude westerly jet stream wind above the mountains splits in two, allowing the southerly branch to bring cyclones originating in the Mediterranean and their heavy rainfall to northern India. In spring, the southerly branch of the jet stream decays, thereby allowing the tropical cyclones that bring monsoon weather to the south of the country to be moved north from their point of origin by the trade winds.

Monsoons, with their heavy rainfall, can be potentially hazardous to humankind, particularly if they coincide with other climatic events. In 1991, a tropical cyclone associated with the onset of the monsoon, generated winds of nearly 150mph and caused a storm surge in the Bay of Bengal that inundated much of this low-lying country and left 120,000 people dead and millions more homeless. Matters were made worse when diseases associated with unburied corpses and waste spread, and the floods were unable to abate as the polluted water was trapped inside many of the dams and embankments build by the Bangladeshis to protect their countryside from flooding. Rescuers struggled to reach the survivors because of the flooding and helicopters proved the most effective means of reaching those who had managed to reach the safety of what little high ground was available.
See: Condensation, Convection, Cyclone, Intertropical Convergence Zone, Jet Stream.

Mother-of-Pearl Cloud
An unusual and rare form of cloud found at high levels in the atmosphere in more northerly latitudes. The cloud is fine, lens-shaped, and gains its name from its iridescent coloring.
See: Cloud, Cloud Classification.

Mountain Breeze
This is part of a localized system of air circulation found in mountain areas. It is associated with deep and wide valleys that face the sun and is most common during the summer. Once the snow cover has disappeared, the valley sides are warmed by solar radiation during the day and, through the process of sensible heating, the transfer of heat from one point (or object) to another by convection or conduction (or both), the air parcel adjacent to the valley side warms and becomes less dense than the cooler air found at the same altitude over the floor of the valley. A circulation pattern is established in which the warmer air rises up the valley slope, expands and cools, and may promote the development of cumulus cloud formation near the mountain peaks, while the air over the central valley floor falls to replace the rising air. This valley breeze system is most prevalent between the mid-morning and sunset, and is replaced by a reverse circulation, known as a mountain breeze, during the night.

On clear nights, the valley walls cool down rapidly as do the adjacent parcels of air. These parcels are now cooler and denser than the air parcel at the same altitude over the valley floor and, as the cooler air descends into the valley bringing cold and gusty winds, the warmer air over the valley rises. Cold air thus builds up in the valley bottom and, if it cools further, fog or low-lying stratus clouds may develop.
See: Conduction, Convection, Cumulus, Radiation, Stratus, Wind.

Mountain Climate
A climate that is primarily influenced by topography and altitude rather than by latitude and distance from the sea. Generally, mountains have lower pressure, temperature, and humidity than areas at lower altitudes.
See: Climatic Zones.

Mud Rain
The name given to rain that contains significant amounts of dust particles within its water droplets. The dust is usually carried into the atmosphere during a duststorm and can be transported over great distances before it falls back to earth.
See: Duststorm, Precipitation.

Nacreous Cloud *(Left)*

Also known as mother-of-pearl cloud, nacreous formations are found in the upper reaches of the Earth's stratosphere at heights of approximately 30 miles. These are rare cloud types, which have a shiny appearance and are believed to consist of either supercooled droplets of water or ice crystals. They are most prevalent during the winter in high latitudes and are seen to best effect when lit by the setting sun.

See: Atmosphere, Cloud, Cloud Classification, Mother-of-Pearl Cloud, Stratosphere.

Neon

One of the component gases of the atmosphere, neon is a colorless, odorless inert gas which comprises approximately 0.00182 percent of the total by volume.

See: Atmosphere.

Nephoscope

This is a device usually used to measure the speed and direction of movement of clouds at both medium and high altitudes. One of the more commonly used instruments, the Besson comb nephoscope, consists of a number of comb-like pointed rods fixed to the top of a long vertical rod. This can be rotated until it appears that the clouds are moving between the points of the comb so that their direction of movement can be identified. Also, if the height of a cloud formation can be found, its speed can be gauged.

See: Cloud, Cloud Classification.

Neutral Air Layer

This is the term given to a layer of air in which a pocket of air that is either descending or ascending has the same temperature and density as its environment. In consequence, a neutral air layer will neither increase or decrease the rate at which a pocket of air either rises or falls.

See: Air Mass, Air Mass Modification.

Nevados

This is a katabatic wind which flows down the mountains into the upper valleys of Ecuador. A nevados, a cold wind, is generated by air cooling in the highlands due to night-time cooling and

contact with ice or snow on the surface, and then blowing downslope.
See: Katabatic Wind, Mountain Breeze, Wind.

Nimbostratus *(Below right)*
This is a dark-gray sheet-like cloud type with a flat base found at low altitudes. Usually associated with unsettled weather, they are the chief precipitation-bring cloud in many frontal systems. Precipitation, often visible as sheets below the base of the cloud, can fall as either rain, sleet, or snow, and is likely to last for some considerable time. If the precipitation is linked to the passage of a warm front, then precipitation is likely to persist for several hours but can vary in intensity depending on the level of development of the front. Shreds of cloud, known as pannus or scud, can develop beneath nimbostratus.
See: Cloud, Cloud Classification, Front, Precipitation, Sleet, Snow.

Nitrogen
This is the most common gas in the Earth's atmosphere, comprising some 78 percent of the total by volume. Colorless and odorless, it is important in its diluting of the atmosphere's oxygen content.
See: Atmosphere.

Nitrogen Dioxide
A gas thought to be contributing to the greenhouse effect and global warming, nitrogen dioxide is caused by air pollution from certain industrial activities, although it does occur naturally when the nitric oxide created by bacteria in the soil combines with oxygen in the atmosphere. Nitrogen oxide can have serious consequences for health, particularly illnesses related to the heart and lungs. If nitrogen dioxide combines with air moisture, it produces nitric acid, which is corrosive and is a key component of acid rain.
See: Atmosphere, Acid Rain, Greenhouse Effect, Global Warming.

Noctilucent Cloud
These are high altitude clouds found in the upper mesosphere at heights above 40 miles and are most common in high latitudes either just after sunset or shortly before sunrise. Such clouds are rare, have a silver-blue appearance when partially illuminated by the sun against a dark night sky,

and probably consist of ice deposits affixed to particles of meteor dust. They have a thin, wave-like appearance.
See: Atmosphere, Cloud, Cloud Classification,, Mesosphere.

Norte
This northerly wind is a continuation of the norther wind from the US which blows through Central America during winter. Norte winds are associated with a sudden and pronounced drop in temperature and strong winds, possibly reaching gale force along coastal areas. On Mexico's northern coast, the norte is accompanied by heavy rain; when it reaches the Pacific coast, the wind has lost much of its moisture content but remains cold. A similar phenomenon is found during winter in eastern Spain, the product of high atmospheric pressure over the country's interior.
See: Norther, Wind.

Norther
Found in the southern US, the norther wind is associated with the passing of a depression. Temperatures drop dramatically and quickly, severe thunderstorms and hail are common, and the wind can reach speeds of between 40 and 60mph. Line squalls are commonly linked to a norther.
See: Depression, Hail, Line Squall, Norte, Thunderstorm.

O

Northern Circuit

This is the path usually followed by depressions as they cross the continental US from west to east, crossing the Great Lakes and St. Lawrence River. Depressions follow this track most frequently during the summer months, although they are weaker and less frequent than at other times of the year.
See: Depression, Southern Circuit.

Northern Hemisphere

The area of the Earth lying to the north of the equator.
See: Southern Hemisphere.

North Pole

The point at the most northerly extremity of the Earth's surface which forms one end of the planet's axis and remains fixed as every other point on the surface rotates around the axis.
See: South Pole

Nor-Easter

This is an intense tropical cyclone which follows a path along the East Coast of the US, particularly between October and May. The system is associated with extremely strong winds blowing onshore, flooding, and damage to property.
See: Cyclone.

Nor-Wester

The names given to particular types of wind. The first is a wind that brings thunderstorms, heavy rain, and hail to the plains of northern India during the April-June hot season, while the second is a hot and dry wind that flows down from the mountains of New Zealands South Island.
See: Fohn, Wind.

Nuclei

The name given to the minute solid or liquid particles on which the deposition or condensation of water droplets take place. These particles are produced by natural forces, such as fires and volcanic eruptions, or can be generated by humankind's activities, such as the burning of fossil fuels. Such nuclei are an important component of cloud formation and two types have been identified: Cloud condensation nuclei and ice-forming nuclei.
See: Cloud, Cloud Condensation Nuclei, Condensation, Deposition, Ice-Forming Nuclei.

Occlusion

Also known as an occluded front, an occlusion develops when a cold front, usually traveling at twice the speed of a warm front, catches up with the warm front. Two types of occlusion have been identified. The first and most common is the cold occlusion which develops when the air in the rear of the advancing cold front is colder than the cool air in advance of the warm front. The cold air pushes under the warm front and its warm air, and also forces the cool air to rise. A warm occlusion occurs when the air behind the advancing cold front is not as cold as the air in advance of the warm front. In this case, the cool air behind the cold front forces the warm air associated with the warm front to rise, but in turn is forced to rise by the colder air in advance of the warm front.
See: Cold Front, Warm Front.

Okta

A measure of cloudiness based on visual analysis of the amount of the sky covered. Usually divided into eighths, with complete cover identified by eight okta.
See: Cloudiness.

Opaco

A word of Italian origin relating to a mountain slope which is orientated toward the poles (north-south) and therefore receives little heat from the sun and is considerably darker and cooler than a slope running east-west.
See: Ubac.

Orographic Cloud

The cloud that develops when a parcel of air undergoes expansional cooling as it rises over a mountain chain.
See: Orographic Lifting, Orographic Rain.

Orographic Lifting

This refers to the process by which a parcel of air is forced to rise and fall by changes in the surface topography across the direction in which the parcel of air is being blown by the prevailing wind. For example, air forced to rise by a mountain chain blocking its path undergoes expansional cooling and, if the increase in its height is sufficient, clouds and rain develop once the air becomes saturated with moisture. The point at

which this begins is known as the lifting condensation level. Once the parcel of air has crossed the mountain barrier, it descends and is heated up by compressional warming, and clouds and rainfall are less likely to develop.
See: Compressional Warming, Condensation, Expansional Cooling, Rain Shadow.

Orographic Rain

The rain that is generated by a parcel of air undergoes expansional cooling as it rises over a mountain chain barrier.
See: Orographic Lifting.

Oxygen

Essential to humankind's occupation of the Earth and essential in combustion, oxygen is a colorless, odorless gas that comprises 20.95 percent of the planet's atmosphere by volume.
See: Atmosphere.

Ozone

A colorless gas smelling of chlorine which is formed by an electric discharge in oxygen. It comprises 0.000007 percent of the Earth's atmosphere by volume. Although it is only a small part of the atmosphere, ozone is an essential gas as it blocks potentially dangerous amount of the ultraviolet radiation generated by the sun.
See: Atmosphere, Greenhouse Effect, Global Warming, Ozone Hole, Ozone Shield, Ozone.

Ozone Hole

Despite being present in the Earth's atmosphere in very small quantities, ozone is vital in reducing the amount of harmful solar ultraviolet radiation, a cause of skin cancer and other illnesses, that reaches the surface of the planet. However, the international scientific community has been expressing concern about the destruction of the ozone shield brought about by humankind's activities.

Chief concern focused on chlorofluorocarbons (CFCs), a group of chemicals developed in the 1920s and used in cooling systems, refrigerators, aerosol sprays, and the manufacture of insulating foams. CFCs are non-reactive in the lower level of the atmosphere (troposphere) but, when they are circulated into the higher stratosphere where the Earth's ozone layer is found, they react with solar

ultraviolet radiation. The ultraviolet radiation causes the CFCs to break down, releasing chlorine, which reacts with ozone, producing chlorine monoxide and oxygen. As the ozone layer is weakened then more ultraviolet radiation reaches the lower atmosphere, causing sickness and warming. It is estimated that every one percent deduction in atmospheric ozone leads to a two percent increase in the intensity of the ultraviolet radiation passing through the ozone layer.

Evidence of possible long-term destruction of the ozone layer was first uncovered in Antarctica in the 1970s and scientists noted the unusually high levels of chlorine monoxide, a byproduct of the breakdown of CFCs by solar ultraviolet radiation. The phenomenon was seasonal, lasting from September to October when the air above Antarctica is cut off from global circulation patterns by a band of strong winds known as the circumpolar vortex. In 1989, scientists discovered an ozone hole developing over the Arctic. In late 1992, representatives of over half of the world's countries met at Copenhagen, Denmark, and agreed to end CFC production by 1996.
See: Atmosphere, Aerosol Gases, Chlorofluorocarbons, Circumpolar Vortex, Global Warming, Greenhouse Effect, Ozone, Ozone Shield, Stratosphere, Troposphere.

Ozone Shield

This is the barrier lying at altitudes of below 30 miles above the Earth that reduces the amount of potentially harmful ultraviolet radiation reaching the surface. Ozone reacts with ultraviolet radiation in two ways. First, ultraviolet radiation is essential in the formation of ozone as it splits oxygen (O2) molecules into two free atoms (O) which are then collide with other oxygen molecules to form ozone (O3). Secondly, ultraviolet radiation also destroys ozone by splitting it into one molecule of oxygen (O2) and one free oxygen atom (O), which can impact with an ozone molecule to form two oxygen molecules. Both reactions therefore greatly reduce the amount of ultraviolet radiation potentially able to reach the surface.
See: Atmosphere, Greenhouse Effect, Global Warming, Ozone Hole, Ozone.

Pacific Air

The name frequently given to a modified form of a maritime Pacific air mass once it has crossed the barrier of the Rockies in the western US. The original maritime Pacific air mass is cool and humid, but is forced to rise by the mountain range. Further cooling leads to condensation, cloud formation, and rainfall as the air mass rises, but once over the barrier, the air, much drier, descends into the Great Plains.
See: Air Mass, Condensation, Orographic Lifting, Orographic Rain, Rain Shadow.

Pampero

This Spanish word refers to an unexpected spell of cold polar air and unsettled weather following in the path of a depression as it tracks across the pampas grasslands of Argentina and Uruguay. The wind blows from a southerly to a westerly direction, is associated with line squalls, and usually brings thunderstorms, heavy rain, and a noticeable drop in temperature as the storm passes.
See: Depression, Line Squall, Thunderstorm.

Papagayo

This is a cold northerly wind which blows across the Mexican interior and is similar to the North American norther and the norte found on the Mexican coast.
See: Norther, Norte.

Parhelion

Also known as a mock sun, this is a phenomenon created at high altitudes by the reflection of light from ice crystals contain in cloud which produces bright spots at the same height above the horizon as the sun. Mock suns are frequently colored and always appear red on the side facing the actual sun.
See: Cloud, Sun.

Partial Drought

Although not accepted internationally, this term is applied in Great Britain to any spell of at least 29 consecutive days in which the average daily rainfall total is less than 2mm.
See: Arid, Absolute Drought.

Penumbra

This is associated with an eclipse and is the partly shaded outer zone surrounding the central dark umbra from which a small amount of light is received.
See: Sun, Umbra.

Perihelion

This refers to the position of the Earth during its orbit around the sun, when it is closest to the sun. The perihelion occurs during winter in the northern hemisphere (about January 3) at which time the Earth is 91 million miles from the sun. The opposite of the perihelion, the date on which the Earth is farthest from the sun (94 million miles), occurs during the northern hemisphere's summer (July 4) and is known as the aphelion.
See: Sun, Season.

Permafrost

This is the layer of ground usually found in high latitudes that remains permanently frozen despite seasonal changes in temperature. It lies at varying depths beneath the layer of soil that thaws and freezes on a seasonal cycle. Permafrost regions are found in the far northern regions of North America and Siberia.
See: Tundra.

Photosphere

This is the visible area of the sun.
See: Plages, Penumbra, Sun, Umbra.

Photosphere *(Illustration page 128/9)*

A key natural process that has played a central role in the evolution and composition of the Earth's atmosphere. It has been estimated that this process, by which plants absorb sunlight, carbon dioxide, and water to generate food and give off oxygen, has had a significant impact on the atmosphere for about two billion years. It also has an important role in the modern world in that the use of carbon dioxide in photosynthesis removes a pollutant from the atmosphere and returns oxygen, a vital gas to humankind's ability to survive on the planet. The destruction of rain forests may have both reduced the amount of carbon dioxide being taken from the atmosphere and, conversely, added to it by the burning of forests. Deforestation is considered by many commentators to make a significant contribution to global warming.
See: Atmosphere, Carbon, Dioxide, Deforestation, Global Warming, Oxygen, Rain forest.

PH Scale

The scale used to test the acidity or alkalinity of a substance by measuring the concentration of hydrogen ions it contains. The scale ranges from 0 to 14, with a reading of seven indicating a neutral substance. Figures above seven indicate increasing alkalinity, while figures below seven show increasing acidity. Pure water has a reading of 0, while rain water, which is naturally acidic as water dissolves with carbon dioxide in the atmosphere, has a pH reading of 5.6. Generally, any rainfall that generates a reading of more than 5.6 is classified as acid rain and recognized as polluted due to humankind's activities.
See: Atmosphere, Acid Rain, Carbon Dioxide.

Pilot Balloon

A method used to gauge the velocity and direction of the flow of winds at high altitude or the height of the cloud base, a pilot balloon is filled with a measured amount of hydrogen which gives a known rate of ascent. Once the balloon is released, it rises through the atmosphere and is tracked by a theodolite. At set intervals, its position is determined and its direction and speed of movement can be calculated, thereby giving the direction and velocity of the wind.
See: Ballonsonde, Radiosonde.

Plages

These are bright areas on the surface of the sun that are usually found in close proximity to the cooler dark sunspots where the temperature may be as much as 2,000°C (3,600°F) cooler than the sun's average temperature. Variations in the intensity of sunspot activity are believed to have some impact on the Earth's climate.
See: Penumbra, Photosphere, Sun, Sunspot, Umbra.

Planetary Circulation

The circulation of air masses around the Earth is dominated by the interaction between large-scale winds and systems of low and high pressure. Chief among the planetary winds are polar easterlies, westerlies, and various trade winds. In general their effects are felt over long distances and can persist for months.
See: Hadley Cell, Horse Latitudes, Intertropical Convergence Zone, Jet Stream, Rossby Waves, Trade Winds.

Plum Rains

The term given by the Japanese to the rains that fall in early summer, plum rains are the product of depressions that develop over continental China and then head across the Yellow Sea, heading for the Japanese mainland. The phenomenon is characterized by dull, overcast skies, frequent bouts of rain, and high humidity. The cloudiness linked to the passage of the plum rains helps to modify the heating that would be normally expected with the onset of summer.
See: Depression, Precipitation.

Pluvial

This means relating to rain.
See: Precipitation.

Polar Amplification

This refers to the fact that significant changes in temperature tend to become greater with latitude.
See: Temperature.

Polar Front

This is the point on the Earth's surface where planetary scale winds meet in both the northern and southern hemisphere. In both cases, its position is at approximately 60 degrees latitude, where polar easterlies meet surface westerlies. The cold and dense air masses flowing from the poles meet the lighter and warmer air masses flowing from the midlatitudes along the polar front. Although the front are not always continuous, they are often the point of origin for large-scale weather systems.
See: Planetary Circulation.

Polar High

This is a type of cold anticyclone that develops in an area where continental polar air masses are prevalent. These anticyclones are created because of the high degree of radiational cooling that is found in high latitudes where there is a great and prolonged degree of snow cover.
See: Air Mass, Anticyclone, Continental Polar Air.

Polar Wind

Found in both the northern and southern hemispheres, these are extreme cold winds which develop over the poles and then blow from these areas of low pressure to areas of high pressure in the planet's temperate zones. In the northern

hemisphere, they flow from the northeast, while in the south they blow from the southeast.
See: Pressure Gradient, Wind.

Poleward Heat Transport

This refers to the movement of heat from lower to higher latitudes, without which the lower latitudes, which receive great insolation, would become increasingly warmer when compared to higher latitudes. There are three methods by which this heat exchange takes place: warmer air masses flow toward higher latitudes, large-scale ocean currents transport heat toward the poles, and storms move heat away from lower latitudes.
See: air Mass, Air Mass Modification, Insolation.

Pollution *(Illustrated pages 131, 134/5)*

Pollution refers to the presence of some substance in the form of a solid, liquid, or gas in the Earth's environment in concentrations that are above what would normally be expected and has a detrimental impact on the natural environment. Humankind's activities impinge on the environment in often complex and little understood ways and nowhere is this more apart than in relation to climate. The rapid industrialization of many countries since the mid-1800s has led to huge amounts of pollutants, chiefly generated by the burning of fossil fuel, being incorporated into the atmosphere. Through processes such as the greenhouse effect, the temperature of the atmosphere has shown a marked rise and this has led to fears of significant climatic change and increases in serious illnesses. Pollution can be controlled by strict policies agreed on an international scale and the damage done to the environment can in all probability be reversed. Often, all that seems lacking is the will and commitment.
See: Aerosols, Climatic Change,
Chlorofluorcarbons, Global Warming,
Greenhouse Effect.

Ponente

The name given to a westerly wind found around the Mediterranean.
See: Mistral, Wind.

Precipitation

Precipitation, which can fall to earth as rain, drizzle, snow, sleet, and hail, is a product of the inter-action between several forces. Under normal conditions, the water droplets and ice crystals that form clouds are so tiny that they will remain suspended in the atmosphere due to the updraft of air within a cloud. Two forces act on a water droplet or ice crystal: gravity, which attempts to force the ice or water to descend at a constant rate, and the friction of the air through which the water droplet or ice crystal has to fall. This increases as they accelerate toward the ground. At some point gravity and air friction forces reach a balance and the water droplet or ice crystal falls at a constant rate known as its terminal velocity.

If the updraft in the cloud is stronger than a water droplet's or ice crystal's terminal velocity, then it will not fall. It will only descend when its terminal velocity is greater that the updraft and to achieve this imbalance the water droplet or ice crystal has to grow.

Two processes have been identified by which water droplets or ice crystals can grow: the collision-coalescence and Bergeron processes. In the case of collision-coalescence, a process most common in warm clouds found in the tropics, it is essential that a cloud's water droplets are of different size because if they were all of the same size, then their terminal velocities would be the same. If they are of different sizes, however, the larger droplets have a greater terminal velocity and are more likely to collide with smaller droplets with a lower terminal velocity. As larger droplets descend, they collide with the smaller droplets and they coalesce (merge), thereby producing a larger droplet. If this process continues, the droplet becomes so large and its terminal velocity so great that it can overcome the updrafts of wind and will fall to earth as rain.

The Bergeron process, especially important in middle and high latitudes, develops in cold clouds containing a mixture of supercooled water droplets and ice crystals. Although the water droplets initially outnumber ice crystals, the reverse becomes true as water droplets vaporize more readily than ice crystals and the vapor is then deposited on the ice crystals. As the ice crystals grow, their terminal velocity increases and they begin to fall, hitting and coalescing with other water droplets and smaller ice crystals.

See: Bergeron Process, Cloud, Condensation, Deposition, Drizzle, Gravity, Hail, Orographic Cloud, Orographic Lifting, Sleet, Snow.

Pressure Gradient

An indication of the rate at which atmospheric pressure changes horizontally between two points on the surface of the Earth. Pressure gradients can be identified on weather charts by the values ascribed to the isobars that connect points of equal pressure. If the isobars are close together, then they indicate that there is a steep pressure gradient running at right angles to the isobars between two recording points. If the isobars are well spaced then the pressure gradient is much less marked. As air flows from areas of high to areas of low pressure, then the steeper the pressure gradient, the stronger the force of the wind.

See: Air Pressure, Isobar, Pressure Tendency, Wind.

Pressure Tendency

A measure of how much the atmospheric pressure has changed at a given point over a set period of time, which indicates whether the pressure is rising or falling. On a weather chart, points subject to equal changes in pressure are connected by isallobars.

See: Isallobar.

Prevailing Wind

The term given to the wind that blows more frequently than any other in a particular direction over a particular point or area on the Earth's surface.

See: Wind.

Psychrometer

A device, consisting of a pair of mercury thermometers, designed to measure the relative humidity of the atmosphere. One of the thermometers (dry-bulb) measures the real air temperature, while the other (wet-bulb) has its bulb enclosed in a muslin wick and the first readings are made after the muslin covering has been soaked in distilled water. Both thermometers are then aerated by turning them at speed until the mercury in the wet-bulb achieves a steady reading. Water soaked into the muslin covering of the wet-bulb thermometer evaporates as the thermometers are aerated by being spun and the temperature falls by cooling. If the air is very dry then the rate of evaporation will be greater than if the air is more humid, and the temperature fall recorded by the wet-bulb thermometer will be greater. Next the value of the temperatures recorded by the two thermometers can be read on a psychrometric chart, which gives the relative humidity as a percentage.

See: Evaporation, Thermometer, Relative Humidity.

Purga

Much the same as the wintry buran, this cold northwesterly wind is found in Siberia, especially in tundra regions, and is usually associated with snow.

See: Buran, Tundra, Wind.

Pyranometer

An instrument for measuring the intensity of solar radiation hitting a horizontal surface, a pyranometer consists of a transparent hemisphere that allows short-wave insolation to pass through it and strike a series of black and white segments within the device's sensor. The white sections reflect back insolation, while the black sections absorb solar radiation. Consequently, because of their thermal and reflective properties, the temperatures of the two different colored segments are affected differently by the same intensity of solar radiation. The difference in temperature between the black and white sections therefore reflects the difference in radiation intensity and their recordings can be transferred to a pen which can record changes on a chart.

See: Insolation, Radiation.

Pyrheliometer

A device for measuring the amount of radiation emanating from the sun.

See: Radiation.

Radiation

This is the process by which a body, such as land or water, gives off radiant energy in the form of heat and is therefore cooled. The sun also gives off radiation, which reaches the Earth and is known as insolation; likewise the Earth is also constantly losing heat into space. During the daytime, the amount of heat gained by the Earth through insolation is greater than that lost through radiation and temperatures rise. At night, radiation given off to space by the Earth is greater than the insolation received from the sun and temperatures fall. Different surfaces also have different abilities to store and give off heat because of their different thermal properties. For example, land gains and loses heat more rapidly than water.
See: Insolation, Sun.

Radiation Fog

Also known as ground fog, this is a low-lying cloud that develops when night-time radiation (heat loss) leads to considerable cooling of an air parcel close to the ground. As the temperature of the air falls, the dew point is reached and condensation begins. Radiation fog is most likely to appear when the air in contact with the ground is moist, the sky lacking in clouds, and winds are light.
See: Condensation, Dew Point, Fog.

Radiosonde *(Illustration page 140)*

A device used to take readings of humidity, pressure, and temperature in the atmosphere up to heights of approximately 20 miles, the radiosonde consists of a balloon to which has been attached various recording instruments and a radio which can transmit the reading back to a ground station.
See: Ballonsonde.

Rain *(Left)*

A type of precipitation consisting of water droplets that condense in the atmosphere, grow in size and then fall to earth.
See: Cloud, Condensation, Drizzle, Precipitation.

Rainbow *(Illustration pages 138/9)*

An optical effect associated with the impact of sunlight on rain and caused by the reflective and refractive properties of the water droplets. Sunlight has a mixture of wavelengths, which are dispersed

by the rain's water droplets to a greater or lesser degree, thereby giving a rainbow its distinctive banded look. The most common rainbow has an angular radius of 42 degrees and has a red outer edge and a violet inner edge. This is known as a primary bow. A secondary bow, the product of sunlight being reflected twice by the water droplets, has an angular radius of 52 degrees and the sequence of colors is reversed. In both cases, the brightness of individual colors depends on the size of the water droplets. The bigger they are, the stronger the colors.
See: Precipitation, Rain, Reflection, Refraction.

Rainfall

The total volume of rain to fall on a particular place over a set period, as measured in a rain gauge.
See: Rain Gauge.

Rain Forest *(Right and pages 142/3)*

Also known as equatorial forest, rain forests are found in the hot, wet, and humid zones around the equator. They play an important role in removing carbon dioxide from the atmosphere, and their clearance by logging companies and farmers

intent on extending the area under cultivation is believed to be contributing to global warming.
See: Carbon Dioxide, Deforestation, Global Warming, Reforestation.

Rain Gauge *(Below)*

A simple device consisting in its most basic form of a funnel fitted to the mouth of a collecting device which is fitted into a larger outer cylinder. Water is unable to escape from the inner vessel and by simply measuring the depth of water contained within it, the total amount of rain to fall at a given point over a set period can be recorded.
See: Precipitation.

Rain Shadow

When an air mass rises to cross a mountain barrier lying across its path, the water vapor it contains condenses due to expansional cooling to produce cloud. If the cooling continues, precipitation begins to fall on the mountain ranges windward slope. Once over the summit, the air parcel descends and warms, thereby reducing its relative humidity and reducing the likelihood of cloud and precipitation. Consequently, the climate on the downwind side of the mountain barrier is significantly drier than that on the windward side. This area is said to be in the rain shadow of the mountain.
See: Condensation, Expansional Cooling, Orographic Lifting, Orographic Rain.

Rain Spell

A term used in Great Britain to identify a period of 15 consecutive days or more in which the daily rainfall total has been 0.2mm or above.
See: Absolute Drought, Drought.

Range of Temperature

The difference between the maximum and minimum temperatures experienced by a particular place over a set time period, this can be measured on daily, monthly, or annual basis. Perhaps of most value for meteorologists is the average temperature range between the hottest and coldest months, known as the mean annual range.
See: Temperature,

Rasputista

This refers to a period of climatic transition in Siberia when winter gives way to warmer weather. Over a period of several weeks, lakes and rivers become free of ice and snow covering the ground melts, leading often to severe floods and thick mud. This period, the Rasputista, ends as temperatures climb, drying out the land.
See: Seasons.

Reforestation *(Illustration page 146/7)*

The process of replanting trees to replace those lost through natural events, such as volcanic eruption or fire, or through humankind's activities, such as logging, land clearance, or damage by pollution. Trees play an important role in removing carbon dioxide from the atmosphere through photosynthesis, thereby helping to reduce the volume of a gas that contributes to global warming. Conversely, the burning of forests, adds carbon dioxide to the atmosphere. When trees are planted in areas where they have never grown before, the process is known as afforestation. The replacement of vegetation, coupled with reductions in manmade pollutants that contribute to the warming of the atmosphere, is likely to reduce the dangers of global warming.
See: Atmosphere, Carbon Dioxide, Deforestation, Global Warming, Photosynthesis.

Reflection

Reflection occurs when light or radiation strikes a particular surface, such as a water droplet or the atmosphere, and bounces back off it. Reflection is important in helping to determine the amount of solar radiation passing into and through the atmosphere, and also has a role to play in such shortlived phenomena as rainbows.
See: Atmosphere, Insolation, Rainbow, Refraction.

Refraction

Refraction occurs when light moving from one transparent substance, such as the atmosphere, to another transparent medium, such as water droplets, is bent. The rays of light bend because the speed of light is faster through the atmosphere than it is through water. Refraction plays an important role in the creation of rainbows.
See: Rainbow

Relative Humidity

This is a means of comparing the true concentration of water vapor in the atmosphere with the concentration of water vapor in the atmosphere if it was saturated. The actual vapor pressure is divided by the saturated water vapor pressure and the result is then multiplied by 100 to give a percentage. If the result is 100 percent, then the relative humidity is 100 percent.
See: Humidity, Specific Humidity.

Ridge of High Pressure

This is the name given to a stretched area of high pressure usually related to an area of high pressure or an anticyclone. On weather charts, it can be recognized by high pressure isobars analogous to the contours on a map which identify a ridge.
See: Anticyclone, Isobar, Trough of Low Pressure.

Rime *(Illustration page 148)*

This type of frost consists of frozen deposits of ice crystals that develop when supercooled water droplets come into contact with exposed surfaces and freeze to them. Rime is usually associated with fog and has a white opaque appearance because numerous air pockets are trapped within it.
See: Fog, Frost.

Roaring Forties

The name given to westerly winds found in the southern hemisphere between 40 and 50 degrees latitude. The absence of land masses in this band means that the winds are both strong and persistent. Depressions follow the path of the Roaring Forties and bring storms and heavy winds.
See: Depression, Planetary Circulation, Trade Winds.

Roll Cloud

A type of cloud formation associated with the development of a thunderstorm, roll cloud forms when the gusty wind and cold air blows away from the center of the storm near ground level, forcing inblowing air to rise. As this rises condensation occurs and clouds develop. As the wind rises it shears (oscillates) and a distinctive layer of cloud is created. If a separate roll of cloud develops in front of the storm due to these processes, it is known as roll cloud.
See: Cloud, Condensation, Shear, Thunderstorm.

Rossby Waves

Named after the meteorologist Carl Rossby, who discovered their existence in the 1930s, these waves are midlatitude westerly winds that flow around the Earth in the troposphere. These winds blow in a wavelike pattern, but the pattern is far from fixed. The number of waves circling the planet varies, as does the wavelength and the amplitude of the waves. Variations in these characteristics are important as they influence the movement of differing air masses from the polar and midlatitude regions, the balance of the heat exchanged between the two, and the development and track of storms.

See: Air Mass, Planetary Circulation, Troposphere.

St. Luke's Summer

In Great Britain, this refers to a period of mild weather supposed to occur on or about St. Luke's day, October 18.
See: St. Martin's Summer, Summer.

St. Martin's Summer

Supposed to occur on or about November 11, St. Martin's day, in Great Britain, this refers to a period of mild weather.

Samoon

Similar to the fohn wind in western Europe, the samoon is a hot and dry wind which blows across Iran from its place of origin in Kurdistan.
See: Fohn, Wind.

Sandstorm *(Illustration pages 150/1)*

A storm usually associated with desert or coastal areas in which strong winds pick up sand particles of considerable size from the ground.
See: Duststorm.

Santa Ana

The name given to a chinook-type wind which is prevalent in southern California during autumn and winter. The origins of the Santa Ana lie in a high-pressure system which is positioned over the Great Basin, from where strong northeast winds drive toward the southwest and the coastal plains. The winds are hot and dry and can gust up to about 90mph. The Santa Ana is notorious for damaging vegetable crops and producing tinder-dry conditions that can encourage the spread of fires.
See: Chinook, Fohn, Zonda.

Saturation Vapor Pressure

This term refers to the maximum water vapor concentration in a set volume of air at a specific temperature. There is a continuous exchange of molecules of water between their vapor and liquid states. When water is evaporating, a greater number of molecules turn to vapor than turn to vapor, and when condensation occurs, a greater number of molecules turn to liquid than turn to vapor. If a balance is established between these two exchanges, the pressure of the water vapor remains constant and is known as the saturation vapor pressure.
See: Condensation, Evaporation, Saturation.

Savanna

Also known as tropical grassland, the world's savanna areas are concentrated in both the northern and southern hemispheres between the equatorial forest and the hot deserts. The climate in the grasslands is typically marked by distinct wet and dry phases. Most of the rain falls during the hot season, when equatorial calms are in place over the grasslands. During the dry season, the savanna regions are under the drying influence of the trade winds.
See: Equatorial Forest, Trade Winds.

Scavenging

This term refers to the dominant natural means by which certain pollutants are removed from the atmosphere's troposphere by precipitation, chiefly rain and snow. Dust particles are bound into snow or trapped by water droplets, and certain gas pollutants dissolve in water droplets. Estimates suggest that over 80 percent of aerosols are removed from the atmosphere by scavenging. However, a note of caution should be made. It is undoubtedly true that scavenging improves the quality of the air, but it only transfers pollutants from the atmosphere to the ground.
See: Aerosols, Atmosphere, Pollution, Precipitation, Snow, Troposphere.

Scotch Mist

This is a form of precipitation consisting of minute water droplets that is similar to mist and drizzle and is found in upland areas. It is the product of thick cloud near the surface and gained its name because it is particularly common in Scotland.
See: Drizzle, Mist.

Scud

The name used by sailors to describe pannus-type cloud. It appears as a secondary cloud located under and either distinct from or linked to other cloud types, such as cumulus and altostratus. It consists of stratus fractus and cumulus fractus, and frequently appears darker that the overlying cloud mass.
See: Altostratus, Cloud, Cloud Classification, Cumulus.

Sea Breeze

This localized wind is a product of the different thermal properties of land and water. During the day, the land warms more quickly than the sea,

thereby establishing a pressure gradient that allows cold air from above the water to move onshore to replace the air that is ascending over the land. Sea breezes develop in the morning, reach a maximum intensity during the afternoon, and die away after sunset. At night-time, the sea breeze is replaced by a land breeze, as the pressure gradient is reversed due to the water cooling down at a slower rate than the land.
See: Breeze, Lake Breeze, Land Breeze.

Sea Fog *(Illustrations page 154)*

This is a type of fog which is produced by advection cooling. When a mass of air moves between two points, its temperature and the concentration of water vapor it holds changes, partly due to the characteristics of the surface over which it flows. If a warm and humid air mass blows over a cold ocean surface, then the lower layers are cooled by advection to saturation and fog may be formed.
See: Advection, Air Mass, Air Mass Modification. Steam Fog.

Seasons

These are periods of the year which are identified by particular climatic conditions. Four seasons have been identified in the world's temperate regions. In the northern hemisphere they are: winter, lasting from December to February; spring. lasting from March to May; summer, lasting from June to August; and autumn, lasting from September to November. In the southern hemisphere, these are reversed so that summer, for example, occurs between December and February.The seasonal succession is caused by the tilt of the Earths axis as it rotates around the sun. The tilt, measured at 23 degrees 27 minutes to a perpendicular to the plane defined by the planets track around the sun, leads to the Earth's orientation to the sun to change continuously as the planet orbits the sun. In conjunction with solar altitude (the angle of the sun 90 degrees or less above the horizon) and length of the day, the changes in orientation influence the amount of solar radiation striking the Earth's surface. The change between the seasons can also be influenced by the thermal properties of land and water. Generally, continental interior's rapid cooling and heating of the surface means that the temperature changes between seasons are greater than maritime regions, which are influenced by oceans which

warm up and cool down at a slower rate than land. Outside temperate regions, the four-season division is more problematic. In the tropics, rainfall is more important than heat in determining the seasons. Usually, two are identified: the wet and dry seasons. Around the poles, the transition between summer and winter is so sudden that it is difficult to identify clear cut periods which could reasonably defined as spring or autumn.

See: Autumn, Fall, Spring, Summer, Winter.

Seistan

This Persian word is the name given to an extremely strong northerly wind capable of speeds in excess of 80mph which blows across the province of Seistan in eastern Iran during the summer.

See: Wind.

Semipermanent Pressure System

Studies of average sea-level pressures indicate that there are areas of the Earth's surface that are dominated by almost permanent low or high pressure. Three particular examples have been identified: subtropical anticyclones, the intertropical convergence zone, and the subpolar lows. Because these features vary in their location on a seasonal basis and their intensity changes through the year, they are termed semipermanent.

See: Intertropical Convergence Zone, Planetary Circulation, Subpolar Low, Subtropical Anticyclones.

Sensible Heating

The term given to the transfer of heat from one substance or one location to another through convection and conduction, or both processes.

See: Conduction, Convection.

Shamal

Found during summer blowing across the basin of the Tigris and Euphrates in Iraq, this northwesterly wind originates from an area of low pressure in northwest India. The wind is associated with duststorms.

See: Duststorm, Etesian Wind, Seistan.

Shear

This term refers to a changes in wind direction or speed over distance.

See: Wind.

Shelf Cloud

Sometimes known as arcus cloud, shelf cloud is associated with the development of a thunderstorm. It is usually found along the edge of a gust front and can be attached and below cumulonimbus cloud. In appearance, shelf cloud is low and elongated with a flat base. It is believed that shelf cloud is created when a parcel of stable, humid, and warm air ascends along the edge of a gust front.

See: Cloud, Cloud Classification, Cumulonimbus, Thunderstorm.

Shower

To the lay person, the term refers to a brief period of rainfall, but to meteorologists it has a much more specific meaning, relating to precipitation that falls from cumulus or cumulonimbus clouds. Generally, showers give an indication of unstable weather conditions as in the case of extreme heating of the surface, which produces strong thermals that can give rise to rain-bearing cumulus congestus and cumulonimbus clouds. Frequently, the strong updrafts of air within these clouds are short-lived. When they occur, they are sufficient to keep large water droplets aloft, but when they subside, the droplets fall suddenly and intensely.

See: Cloud, Cloud Classification, Cumulonimbus, Cumulus.

Simoon

Found in the Sahara and Arabian desert during spring and summer, the simoon is a hot and dry wind that carries large amount of sand.

See: Sandstorm.

Sirocco

This is a wind found in the Mediterranean, especially in North Africa, Italy, and Sicily. It originates within the Sahara as a hot and dry wind laden with dust. As it moves away from its point of origin, it crosses the barrier of mountains running along the North African coast, the sirocco descends and becomes even more dry and hot. As the wind then tracks across the Mediterranean Sea, it gains moisture and reaches the opposite shore as a hot and humid wind. A sirocco wind tends to form in front of the depressions that cross the Mediterranean from west to east on a regular basis. It is particularly common during spring, when the depressions are both frequent and well developed, and the Sahara is hot. Under these

circumstances, the sirocco is likely to last two days before being superseded by cooler northerly wind following in the path of the depressions.
See: Chili, Gibli, Khamsin, Leveche.

Sleet
A term that has two distinct meanings. In the US, it is used to refer to either the pellets of ice that comprise frozen rain drops, or snow which has been partly melted and then refrozen as it has fallen through a layer of cold air close to the ground. In Great Britain, sleet refers to a form of precipitation that consists of a mix of snow and rain.
See: Precipitation, Snow.

Smog *(Above)*
Derived from smoke and fog, smog is a fog containing smoke generated by the combustion of fossil fuels or other pollutants and is commonly found lying over industrial areas or large cities. One particularly troublesome form is known as photochemical smog. This is generated during urban morning and evening rush hours when traffic stopped in jams spews out large volumes of oxides of nitrogen. These react with sunlight to create smog containing a mixture of aerosols and gases that can cause breathing problems as well as irritation. Episodes of smog can be longer lasting and more dangerous, if they occur under certain circumstance. For example, cities located in coastal regions in the lee of a mountain layer are particularly susceptible. Descending warm air is prevented from reaching ground level by cold air blowing onshore, thereby preventing mixing that might dilute the smog.
See: Aerosols, Temperature Inversion.

Smokes
Similar to the cacimbo, this is a type of mist found in the mornings and evenings of the dry season along the west coast of equatorial Africa.
See: Cacimbo, Mist.

Snow *(Illustration pages 158/9)*
This is a type of precipitation formed when the temperature of water vapor in the atmosphere drops below freezing point. Snow itself consists of a number of ice crystals that have coalesced to form flakes. Snowflakes have no set form and they can be needles, columns, stars, or flat plates, depending of the temperature and the concentration of water vapor. Their size depends on temperature and water vapor availability. At relatively high temperatures, snowflakes are moist and more readily stick together than at colder temperatures. With regard to water vapor, at low temperatures

concentrations of water vapor are low so snowflakes are relatively small.
See: Precipitation.

Snowbelt

An area where the likelihood of snow is high, particularly lying downwind of large expanses of water.
See: Lake-Effect Snow, Snow.

Snowdrift

This refers to the buildup of snow often in a sheltered spot where the speed of the wind carrying the snow falls below that required to transport it.
See: Snow, Wind.

Snowfield

Any large area which has a permanent covering of snow.
See: Snow.

Snow Gauge

A device for recording the depth of snow that has fallen at a particular point over a set period of time. Frequently, they are rain gauges modified for snow collection. Although approximately 10cm (4in) of snow fall is roughly equivalent to 1cm (0.4in) of rainfall, the matter is somewhat complicated due to differences in the nature of the snow itself.
See: Rainfall, Rain Gauge, Snow.

Snow Grains

These are a type of frozen precipitation which consists of particles of white ice with a diameter of under 1mm (0.04in). Snow grains are produced by the same process that creates drizzle.
See: Drizzle, Snow, Snow Pellets.

Snowline

This is the variable line on the side of a mountain or along a slope which represents the lower limit of permanent snow cover. The height at which the snowline occurs varies with local conditions, such as summer temperatures, which determine the degree of seasonal melting, and exposure, but the snowline's altitude generally falls away from the warmer tropics toward the colder polar regions. One other factor that is important is the volume of snow that falls in winter. Generally, the snowline in a dry region is higher than in a wet one.
See: Snow.

Snow Pellets

Essentially a type of frozen precipitation that differs from snow grains, particles of white ice, on account of their size. Snow grains have a diameter of less than imm (0.04in), whereas snow pellets have a diameter of 2–5mm (0.08–0.2in).
See: Snow, Snow Grains.

Soft Hail

Similar to snow and often a herald of snowfall, soft hail consists of grains of slightly melted hail.
See: Graupel, Hail.

Solano

Responsible for blowing rain-bearing clouds over southeastern Spain and the Straits of Gibraltar, the solano is an easterly wind.
See: Levanter, Wind.

Solar Constant

This is a convenient measure of the average amount of solar energy that reaches the Earth. It is defined as the rate at which solar radiation hits a surface positioned at the top of the Earth's atmosphere and is perpendicular to the sun's rays when the Earth is at an average distance from the sun. However, while this average is of some value, the solar energy reaching the Earth does vary. In particular, when the planet is at the perihelion (nearest to the sun) the value is higher by 6.7 percent than when the Earth is at its aphelion (farthest from the sun).
See: Aphelion, Perihelion, Sun.

Solar Eclipse *(Illustrated pages 160 and 161)*

This occurs when the light emanating from the sun is blocked from reaching the Earth by the moon which casts its shadow on the Earth. A total eclipse occurs when the sun is completely or very nearly totally obscured, while a partial eclipse takes place when the shadow of the moon does nor reach the Earth. A lunar eclipse occurs when the light from the moon is obscured by the Earth as it passes between the moon and the sun, throwing its shadow over the moon.
See: Penumbra, Umbra, Sun.

Solar Irradiance

This is a measure of the sun's total output of radiation.
See: Sun.

Solstice

There are two solstices, days which mark the middle of summer and winter. They occur when the sun is vertically above the point representing its farthest distance north or south of the equator. In the northern hemisphere, the summer solstice is on about June 21, when the sun is vertically over the Tropic of Cancer, while the winter solstice occurs on about December 22, when the sun lies vertically over the Tropic of Capricorn. In the southern hemisphere the dates are reversed, for example, December 22, is the summer solstice. Generally, the summer solstice is marked by the longest day and shortest night and the winter solstice is marked by the longest night and shortest day in both hemispheres.
See: Northern Hemisphere, Southern Hemisphere.

Soundings

A general term which is used by meteorologists to identify continuous readings, such as temperature, pressure, and humidity, taken at altitude.
See: Meteorology.

Southerly Buster

This is a sudden burst of cold air advancing from the south and moving over southern and southeastern Australia and usually following a trough of low pressure stretching from Antarctica. It is most frequent in spring and summer. The onset of a southerly buster is heralded by strong and cold winds which are often dust laden and accompanied by thunderstorms. Temperature drops are sudden and dramatic.
See: Pampero, Trough of High Pressure.

Southern Circuit

The route taken by depressions as they track from west to east across the continental United States, particularly in winter, when depressions may reach as far south as the Gulf of Mexico.
See: Depression, Northern Circuit.

Southern Hemisphere

That area of the Earth lying to the south of the equator.
See: Northern Hemisphere.

South Pole

The point at the southern extremity of the Earth, marking one end of the planet's axis and which remains stationary while all other points, except the North Pole, rotated around the axis.
See: North Pole.

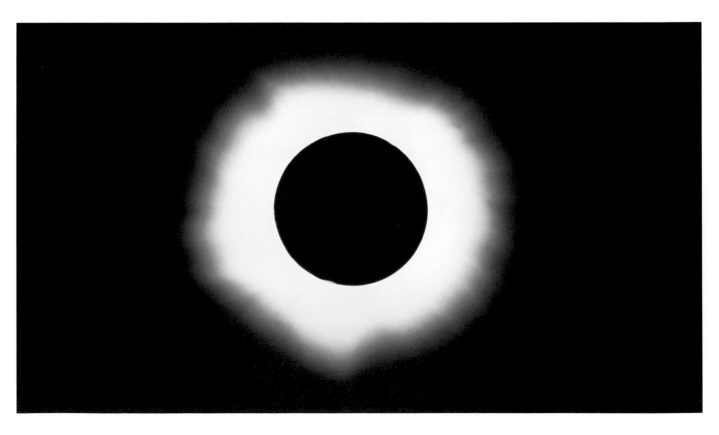

Specific Humidity

This is defined as the ratio of the mass of water vapor to the mass of humid air which contains the water vapor. The result is expressed as grams of water vapor per kilogram of humid air.
See: Absolute Humidity, Relative Humidity.

Sporer Minimum

The name given to a period of cold climate experienced by Western Europe between 1450 and 1550 thought to be related to an increase in the sunspot activity.
See: Maunder Minimum, Sunspot.

Spring

In the northern hemisphere, this refers to the period between March and May during which temperatures begin to rise. In the southern hemisphere, spring falls between September and November.
See: Northern Hemisphere, Seasons, Southern Hemisphere.

Squall

This is a sudden, violent wind of short duration.
See: Squall Line.

Squall Line

This is a band of violent thunderstorms often associated with the decay of a supercell storm. When this begins, it throws out a surge of cold wind. This blast forces warm air along a long line to create an elongated active core. The onset of a squall line is heralded by shelf or roll cloud.
See: Roll Cloud, Shelf Cloud, Supercell Thunderstorm, Squall.

Stable Air Layer

The degree of stability in the atmosphere is determined by comparing the change in temperature of a parcel of air as it rises or falls against the temperature changes occurring in the air through which the parcel of air is ascending or descending. Generally, air warmer than its surroundings rises, while air colder than its surroundings falls. Within a stable air layer, rising air becomes cooler and denser than the surrounding air, while falling air becomes warmer and less dense that its surroundings. Consequently, the cooling warm air will eventually descend back to its original position and the falling air that is warming will begin to ascend to its original position.
See: Adiabatic Lapse Rate, Unstable Air Layer.

Standing Front

This is a front that is fixed over a particular position. There are several reasons why a front might

lose its momentum, including a mountain barrier blocking its path or when a front reaches a thermal boundary between, for example, a snow-bound region and one where the surface lacks such cover.
See: Front.

Steam Fog

This is a form of fog which develops when a mass of cold and dry air blows over a body of ice-free water. The lower level of the cold air is modified by the comparatively warm water which causes the lower layer of cold air to become more humid and warmer that the layers of air immediately above. The mixing of this higher cold, dry air and the mild, humid air promotes saturation and fog develops. The fog appears in streams which resemble smoke or steam, hence the name.
See: Arctic Smoke, Fog.

Stevenson Screen *(Right)*

It is standard practice at weather stations to record shade temperatures, so it is crucial that thermometers are shaded from direct sunlight. One of the simplest devices to facilitate this is known as the Stevenson screen, which consists of a louvered box to aid the free circulation of the air. The box is painted white and stands on legs some four feet above the ground. It has a double roof and one side is hinged so that observers have easy access to the instruments inside, including wet- and dry-bulb thermometers and maximum and minimum thermometers. All records reflect the temperature at four feet off the ground. Larger boxes might contain a thermograph and a hygrograph.
See: Hygrograph, Thermograph, Thermometer.

Storm *(Illustrated pages 164/5)*

A violent climatic event accompanied by high winds and heavy rainfall. On the Beaufort scale, a storm is defined as a Force 10 wind on land and is expected to cause significant damage to property.
See: Beaufort Scale, Supercell Thunderstorm.

Storm Surge

A potential dangerous and usually sudden rise in the height of the sea, usually the product of strong onshore winds and low air pressure, that can be blown onshore to cause flooding and property damage.
See: Cyclone, Hurricane, Storm.

Stowed Winds

These are winds that are blocked by a barrier, such as a range of mountains, and are forced to blow through gaps, thereby increasing their speed.
See: Wind.

Stratiform Cloud

The name given to clouds at low, middle, and high altitudes that exhibit a marked lateral spread.
See: Cloud, Cloud Classification, Cumuliform Cloud.

Stratocumulus

This is a form of low-altitude cloud with a gray or white color and a sheet-like appearance. Stratocumulus have a very distinctive look consisting of a number of pillows or heaps. Two mechanisms have been identified in the formation of stratocumulus. The first occurs when cumulus clouds ascend from the surface until they meet a temperature inversion which forces the clouds to flatten and spread sideways. The second method occurs when sheets of stratus are fractured by convection. Stratocumulus clouds are indicative of clear and settled conditions, although they may produce light showers.
See: Cloud, Cloud Classification, Cumulus, Stratus.

Stratosphere

The name given to that zone of the Earth's atmosphere that lies between the troposphere and the mesosphere. Beginning about 6.5 miles above the ground, it is separated from the troposphere by the tropopause, and reaches a height of about 30 miles, where it is separated from the mesosphere by the stratopause. Temperature changes with height in the stratosphere are not constant. Up to a height of about 12 miles above the surface, temperature remains constant, it then rises until the stratopause is reached. Scientific evidence suggests that, because there is little air circulation between the troposphere and the stratosphere, pollutants, particularly aerosols, are building up in the latter and damaging the protective ozone layer.
See: Aerosols, Atmosphere, Tropopause, Ozone Layer.

Stratus *(Illustrated page 166)*

The name given to a form of low, thin cloud which has a gray coloring and is identical to fog. Stratus

are generated by a parcel of moist air rising slowly or by nearly saturated air is carried by a gently breeze over a cold surface. Precipitation from stratus clouds is often limited to drizzle or an insignificant fall of snow. If a stratus cloud is broken into smaller pieces it is known as stratus fractus.
See: Cloud, Cloud Classification, Fog.

Streamline
The path taken by an air mass as it moves from one point to another.
See: Air Mass, Air Mass Modification

Sublimation
This refers to the process in which water changes from a solid to a vapor without going through the transitional phase of being a liquid.
See: Deposition.

Subpolar Low
Two subpolar areas of low pressure, known as the Icelandic Low located over the North Atlantic and the Aleutian Low over the North Pacific, are found in the northern hemisphere. These lows mark the convergence of planetary winds, the midlatitude southwesterly and the polar northeasterly winds, which bring together differing air masses. In the southern hemisphere, there is a band of almost continuous low pressure where the midlatitude northwesterly and polar southeasterly winds meet. Where the winds converge in both hemispheres is known as the polar front.
See: Northern Hemisphere, Planetary Circulation, Polar Front, Southern Hemisphere, Wind.

Subpolar Region
The name given to the zone sandwiched on its poleward edge by areas of tundra and by the side closer to the equator by cooler temperate or desert-type regions. In northern latitudes in the northern hemisphere, the subpolar region is almost always linked to areas of coniferous forests. Typically, winters in subpolar regions are long and extremely cold, while summers are brief. Rainfall totals are usually low.
See: Temperate Zone, Tundra.

Subtropical Anticyclones
These are a type of semipermanent high-pressure system found close to the subtropical latitudes 30 degrees north and south of the equator, and are

especially prevalent in the North and South Pacific, North and South Atlantic, and the Indian Ocean. These features exhibit extreme vertical development, rising from the surface into the tropopause, and have a major impact on weather and climate. The anticyclones are characterized by subsiding stable air which stretches outward toward their eastern margins and by less stable air on their western fringes. Both have a significant impact. On the eastern margins, the descending stable air is affected by compressional warming, which is associated with warm conditions and low humidity. Consequently, many of the world's major deserts are located to the east of the subtropical anticyclones. Conversely, on the western margins the less stable air is affected less by subsidence and storms are more frequent. Subtropical anticyclones have a very shallow pressure gradient across their centers and horizontal winds at the surface are generally light and calms are frequent. These calms were well known to sailors, who named the area the horse latitudes. In the northern hemisphere, winds blow out from the edges of the subtropical anticyclones in a clockwise direction and are known as the midlatitude westerlies to the north of the horse latitudes (center of the subtropical anticyclone) and as the trade winds on the southern edge. In the southern hemisphere, the situation is reversed, with the winds blowing outward in an anticlockwise direction, with midlatitude westerlies blowing from the southern margins of the subtropical anticyclones and the trade winds blowing to the north. The trade winds from the northern and southern hemispheres' subtropical anticyclones meet at the intertropical convergence zone, which is associated with an area of calm known as the doldrums, while the westerlies meet polar easterlies along the polar fronts.

See: Anticyclone, Compressional Warming, Horse Latitudes, Intertropical Convergence Zone, Midlatitude Westerlies, Planetary Circulation, Polar Front, Pressure Gradient, Trade Wind, Tropopause.

Sulfur Oxide

Huge volumes of sulfur oxide are ejected into the atmosphere during volcanic eruptions. Once in the atmosphere, sulfur oxide can produce sulfuric acid and sulfate particles, collectively known as sulfurous aerosols. These tend to congregate in the stratosphere, where they have two implications for the Earth's climate. First, they absorb solar radiation, thereby warming the stratosphere and preventing the radiation from reaching the troposphere. Second, sulfurous aerosols also reflect some of the insolation back into space, again reducing the temperature of the atmosphere.

See: Atmosphere, Insolation, Stratosphere, Troposphere, Volcano.

Sumatra

A type of weather system located in the Malacca strait, this a squall, usually most common during the night in the local monsoon season. It is associated with sudden changes in the direction of wind from southerly to westerly, heavy cumulonimbus cloud, dramatic drops in temperature, violent thunderstorms and heavy rainfall.

See: Cumulonimbus, Monsoon, Thunderstorm.

Summer

One of four seasons. In the northern hemisphere, summer falls between June and August, while in the southern hemisphere summer occurs from December through to February.
See: Seasons.

Sun *(Illustration page 168/9)*

The closest star to the Earth, the sun consists primarily of hydrogen and helium and is characterized by extremely high temperatures of over 20 million°C which are created by a continuous nuclear fission reaction within the sun's core. While some of the energy produced by the nuclear reaction is reused in the reaction between the hydrogen and helium, the remainder reaches the surface of the star to be radiated off into space. The Earth receives some of this radiation, estimated to be one two-billionth of the total. Known as insolation, it has three components: infrared radiation (46 percent), sunlight (45 percent), and ultraviolet radiation (9 percent). Insolation plays a key role in the Earth's climate and weather patterns.
See: Atmosphere, Insolation, Sunspot.

Sundogs

Another term for parhelia, the two spots of bright light which appear at either side of the sun and are the product of the refraction of sunlight by ice crystals in the Earth's atmosphere. Also known as a mock sun.
See: Atmosphere, Parhelion, Refraction.

Sunshine

This is the light the Earth receives from the sun consisting of those wavelengths of solar radiation between those associated with infrared and ultraviolet light.
See: Sunlight, Sunshine Recorder.

Sunshine Recorder

A simple device used to record the amount of sunshine over a given period, usually a day. The recorder consists of a glass lens fitted in the center of a framework into which a strip of graduated card has been fixed. As the sun shines, the image of the sun on the lens generates sufficient heat to burn a trace on the card. The length of the burn track indicates the number of hours of sunlight over the period in question.
See: Sun.

Sunspot

This is the name given to the large dark spots that periodically develop on the surface of the sun. A typical sunspot consists of a central zone, known as the umbra, which is darker that the outer ring, the penumbra. It is believed that there is a direct link between increases and decreases in sunspot activity and variations in the Earth climate. Generally, increases in sunspot activity and the associated increase in plages, increases the amount of solar irradiance, thereby possibly increasing the amount of solar radiation available to warm the Earth.
See: Maunder Minimum, Plages, Solar Irradiance, Sporer Minimum, Sun.

Supercell Thunderstorm

This is a large weather system that has a long life and is often associated with the development of tornadoes. These supercells are characterized by extremely powerful updraft winds capable of exceeding 150mph. The genesis of a tornado in a supercell thunderstorm starts with the violent interaction between the violent updrafts and the large-scale horizontal wind propelling the storm. The horizontal wind in such a system tends to rise with height above the ground and also veers anticlockwise with altitude. The shear of the wind with altitude encourages the air to rotate about the horizontal in a rolling motion and when this rolling air encounters the violent updraft it is forced from the horizontal to the vertical. This is known as a mesocyclone and if it develops further it is classified as a tornado.
See: Shear, Tornado.

Synoptic Chart

Also known as a weather chart, a synoptic chart contains details relating to certain weather measurements over a particular area at a particular time. The data displayed on a chart is gathered from numerous weather stations dotted about the area covered by the map and includes such information as variations in atmospheric pressure, with points of equal pressure being linked by isobars usually drawn at two or four millibar intervals, the position of areas of high and low pressure, the location of anticyclones, fronts, and depressions.
See: Anticyclone, Depression, Front, Isobar, Millibar.

Taiga

This refers to the zone of coniferous forest that stretches across much of the sub-Arctic latitudes of the North American, European, and Asian continents and lies between the tundra region to the north and the steppe lands to the south.
See: Tundra.

Temperate Zone

This is the term applied to those parts of the Earth in both the northern and southern hemispheres which lie between the hotter zones around the equator and the colder Arctic and Antarctic circles. Generally, the sun is never directly overhead in the two temperate zones, but insolation shows marked seasonal variations, which, in turn, leads to distinctive seasons, chiefly summer and winter. Such climates are usually warm and dry in summer, and cold and wet in winter. The temperate zone in the midlatitude regions of the southern hemisphere is better defined, while the term is only really applicable to the various continents' western coastal regions.
See: Climatic Zones, Insolation, Northern Hemisphere, Southern Hemisphere.

Temperature

Any substance consists of many atoms or molecules that are in continuous, rapid, and random motion. This is called kinetic energy, and temperature is directly proportional to the average kinetic energy of the individual atoms or molecules.
See: Centigrade Scale, Fahrenheit Scale.

Temperature Gradient

The variation in temperature between two points. Gradients can be horizontal as, for example, between the Earth tropics and polar regions, or vertical as, for example, in the troposphere, where temperature falls with height.
See: Atmosphere, Troposphere.

Temporales

These are the strong southwesterly winds found along the Pacific coast of Central America during the summer. They are hot and humid, and bring weather similar to the monsoons of southern Asia.
See: Monsoon.

Terral

The name given to a type of land breeze found along the west coasts of Chile and Peru in South America.
See: Land Breeze, Virazon.

Thaw

The point at which frost, ice, and snow begin to melt due to a rise in temperature above freezing point.
See: Frost, Ice, Snow.

Thermal Belt

This is the belt usually found about 300ft above the bottom of valleys in the Appalachian mountains in the eastern United States, which is not generally affected by frost as temperatures are markedly higher than those found along the valley floor.
See: Frost.

Thermal Equator

Sometimes also known as the heat equator, this is a line running around the Earth connecting points that have the highest mean average temperature over a specified period. It moves north during the summer season in the northern hemisphere, but tracks southward during the southern hemisphere's summer. As mean average temperatures are not solely influenced by the positional relationship between the Earth and the sun, but also reflect the impact of land and sea and the influence of ocean currents, which have a greater impact in the northern hemisphere, the thermal equator is always positioned north of the equator.
See: Northern Hemisphere, Southern Hemisphere.

Thermal Wind

Horizontal wind speed between two points tends to rise as altitude above the ground increases and the impact of the friction layer decreases chiefly because the horizontal pressure gradient, that is the contrast in temperature between a hot and cold area, between two points is usually greater at higher altitudes. Thermals winds, along with coriolis force, are the two components of geostrophic winds.
See: Coriolis Force, Geostrophic Wind, Pressure Gradient, Wind.

Thermograph

This is a type of thermometer which provides a continuous, self-recorded trace of changes in temperature over a set period of time. Recordings are made on a thermogram, which is attached to a clockwork rotating drum.
See: Thermometer.

Thermometer *(Below)*

A recording device, frequently consisting of a glass tube atop a bulb of mercury or alcohol, which records changes in temperature. As temperature rises so too does the level of the mercury or alcohol in the tube and the trace falls as temperature drops. Temperatures are recorded on either the Fahrenheit or Centigrade Scale. Credit for inventing the first thermometer has been given to Galileo Galilei in the late 1600s.
See: Centigrade Scale, Fahrenheit Scale

Thermosphere

This is the outer zone of the Earth's atmosphere, situated above the mesopause, which lies at approximately 50 miles above the surface. Within the thermosphere, temperature initially shows lit-

tle change with altitude, but above about 63 miles temperatures rise rapidly, reaching over 80°C (176°F) at its outer edge, approximately 75 miles above the ground.
See: Atmosphere, Mesopause.

Thunder

The loud rumbling noise linked to lighting in a thunderstorm. Lightning brings about a sudden and very intense heating of the air along the path that it follows, typically in excess of 25,000°C (45,032°F). This causes the air along that path to expand violently, thereby generating the sound waves.
See: Lightning, Thunderstorm.

Thunderstorm *(see illustration page 170)*

A thunderstorm is a weather system that is comparatively shortlived and impacts on only a small area. Such systems originate when descending cold air forces warm air to ascend. If this convection is powerful enough, then the warm air may rise high into the troposphere. Scientists have identified three stages in the growth and decay of a thunderstorm. In the first, cumulus, stage, the clouds associated with a thunderstorm begin to grow both vertically (up to heights of over 30,000ft) and horizontally (over distances of between six to nine miles). Cumulus clouds develop because of either free convection, the product of the heating of the surface by solar radiation, or forced convection due to orographic or frontal lifting of an air mass. Thunderstorms usually develop when free convection is reinforced by forced convection. As the warm air rises, it reaches its convective condensation level, condensation begins and clouds develop. Initially, the air cools at the dry adiabatic lapse rate, but once condensation has occurred, the air continues to rise but cools at the moist adiabatic lapse rate. The cumulus clouds associated with the thunderstorm begin to billow upwards giving rise to cumulonimbus, the cloud formation identified with a thunderstorm. However, precipitation does not begin because the strong updraft prevents water droplets or ice crystals falling. The second, mature, stage conventionally begins when the first precipitation from a thunderstorm reaches the ground, once the updraft is no longer sufficiently strong to keep water droplets or ice crystals aloft, and can last for approximately 30 minutes. The fall of precipitation creates a pronounced downdraft of air within the thunderstorm. This draws in unsaturated air from the edges of the thunderstorm, which vaporizes a fraction of the storm's water droplets and ice crystals, thereby weakening the uplift and reinforcing the downdraft. Thunderstorm activity, heavy rain, thunder and lightning, and strong, gusty surface winds, reach their maximum intensity toward the end of the mature stage. The third, dissipating, stage develops when the subsiding downdraft of air within the thunderstorm becomes much stronger than the updraft. Consequently, the flow of moisture provided by the rising air declines until compressional heating warms the descending air, leading to a decrease in its relative humidity and vaporizing the cloud.

See: Adiabatic Lapse Rate, Compressional Warming, Cloud Classification, Convection Cumulonimbus, Cumulus, Front, Precipitation, Relative Humidity, Tornado, Troposphere.

Timber Line

The line at various altitudes in the mid- and low latitudes beyond which tree cover is absent. The line varies because of several factors. Generally, the timber line is higher in tropical zones than temperate zones, and is generally higher in areas which are protected from cold winds and receive greater amounts of sunshine.

See: Climatic Zones, Sunshine, Temperate Zone, Tropics.

Tornado *(see illustrations pp174/5 and 176/7)*

Tornadoes, potentially the most dangerous of weather systems, are usually produced by a particularly powerful storm, known as a supercell thunderstorm. Supercells are characterized by powerful updrafts of air of often more than 150mph and can last for long periods, thereby potentially giving rise to a series of tornadoes. Fundamentally, a tornado comprises a tapering, whirling column of air that has made contact with the ground and turns rapidly about its vertical axis. The cloud formation most identified with a tornado is known as funnel cloud, which forms because of the extremely sharp pressure gradient that is found between the outer edge and core of a tornado. When humid air is drawn into a

developing tornado, it expands and cools below dew point, condensation takes place, and clouds form. Tornado development from a supercell thunderstorm begins when the horizontal wind blowing into a thunderstorm begins to show signs of increasing speed and turns clockwise with altitude. The wind begins to rotate and, when it reacts with the updraft of air within the supercell, it starts to flow vertically. For a tornado to develop from a supercell thunderstorm, the column of rotating air must become narrower and, as it does so, wind speeds increase greatly. If a tornado does develop, then it is most likely to be found close to the updraft within the supercell but usually toward it trailing edge. Tornadoes vary in both intensity and duration. Powerful ones have reaches speeds of more than 300mph, have a duration of over two hours and can leave a trail of destruction over 100 miles long and many hundreds of yards wide; less powerful tornadoes may have wind speeds of under 110mph, can last for less than five minutes, and leave an area of destruction less than a mile long and 300 yards wide. The F-scale has been devised to categorize tornadoes on the basis of wind speed: F-scales 0 and 1 refer to weak tornadoes, scales two and three indicate strong systems, while four and five identify violent tornadoes. It has been estimated that in the United States on an annual basis some 79 percent of tornadoes fall into the weak range, 20 percent are strong, and only one percent are identified as violent. Nevertheless, tornadoes are destructive. The worst recorded tornado in the United States occurred on March 18, 1925. Parts of three states, Missouri, Illinois, and Indiana, were affected by what may have been a series of tornadoes rather than a single system. An average wind speed of nearly 75mph was recorded for an event that lasted for over three hours and caused damage over an area of close to 220 miles. When clearing-up operations were concluded, investigators estimated that nearly 700 people had died and that 2,000 had been injured. More than 10,000 people were also left homeless. Tornadoes cause damage not solely because of the high winds associated with them and the dangerous debris that they can carry. Another danger lies in the strong updraft of air found at the center of any storm, which is capable of lifting a trailer or some other large object off the ground. Tornadoes are not necessarily confined to

a particular region, although some areas are more prone to them than others. In the United States, for example, all the states have suffered from tornadoes, but they are generally concentrated within a zone stretching from the east of Texas, through Oklahoma, Kansas, and areas of Nebraska — a zone known as tornado alley. Tornadoes in these areas also exhibit certain seasonal and daily characteristics. In general, most form in the warmest part of the day and over 75 percent develop between March and July. Although annual totals vary in the United States, an average of a little over 800 tornadoes per year were identified between 1961 and 1990.

See: Condensation, Dew Point, Funnel Cloud, Scale, Thunderstorm.

Trade Wind

The name given to the regular, planet-scale winds found in the Earth's tropical regions. In the northern hemisphere, the trade winds are northwesterlies, blowing in a clockwise direction away from the southern edges of subtropical anticyclones, while in the southern hemisphere, they are southwesterlies, blowing in an anticlockwise direction away from the northern margins of subtropical anticyclones. The trade winds from both hemispheres meet along the intertropical convergence zone, where they converge to produce the wide belt of light winds known as the doldrums.

See: Anticyclone, Doldrums, Intertropical Convergence Zone.

Tramontana

The name used to describe a cool and dry northerly wind found around the Mediterranean.

See: Wind

Transpiration

Part of the hydrological cycle, this is the method by which vegetation removes moisture from the soil via root systems and then transfers it back to the atmosphere as water vapor via its leaves.

See: Evapotranspiration, Hydrological Cycle.

Tropical Depression

A term indicating an early stage in the development of a hurricane. Once a tropical disturbance has grown and its winds reach a speed of 23–63mph, it is renamed a tropical depression.

Systems that exceed 63mph are known as a tropical storm, and when they break through the 73mph barrier they are officially declared a hurricane.
See: Tropical Disturbance, Hurricane.

Tropical Disturbance

This is the name given to the groups of thunderstorms, identified as areas of low pressure at the surface, that develop over tropical oceans where sustained convection is taking place and which may grow into fully-fledged hurricanes.
See: Hurricane, Thunderstorm.

Tropical Humid Zone *(Left)*

This refers to a fractured climatic zone that lies between the Tropics of Cancer and Capricorn. Average monthly temperatures are high and vary little throughout the year due to the intensity of solar radiation and length of day, both of which also vary little throughout the year. Although temperature patterns vary little in such regions, tropical humid climates can be subdivided on the basis of their precipitation distribution. Rainfall totals are high in both subdivisions. Tropical wet-and-dry climates, as found in savanna regions of much of central Africa and South America, have wet summers and dry winter, while tropical wet climates, as identified with the rain forests of Brazil's Amazon basin, have no marked dry season, with precipitation evenly distributed throughout the year for the most part. Tropical humid climates are identified by the letter A on map of climatic zones, while tropical wet are coded Ar and tropical wet-and-dry are marked by Aw.
See: Climatic Zones, Rainforest, Savanna, Tropics.

Tropics

The zone of the Earth lying between the Tropic of Cancer in the northern hemisphere and the Tropic of Capricorn in the southern hemisphere.
See: Climatic Zones, Tropical Humid Zone.

Tropopause

Part of the atmosphere, this marks the boundary between the troposphere and the stratosphere.
See: Atmosphere, Stratosphere, Troposphere.

Troposphere

This is the lowest level of the Earth atmosphere, rising from ground level to heights of 10 miles above the equator and 3.7 miles above the poles. Generally, average temperatures fall with altitude within the troposphere at a rate of 6.5°C per 1,000 meters (3.5°F per 1,000 feet). The troposphere is where most of the Earth's weather takes place.
See: Atmosphere.

Trough of Low Pressure

The term used to describe an elongated area of low pressure, frequently linked to a depression, where a front might be found.
See: Depression, Front, Ridge of High Pressure.

Tundra

Found only in the northern hemisphere, this zone is found in the north of North America, Europe, and Asia and is situated between the Arctic circle and coniferous forest regions. Winters are prolonged and extremely severe, while summers are short but warm. Trees are few and stunted, although mosses and lichen may appear in summer. The lower layer of the soil are permanently frozen, and summer thawing of the upper soil layers may lead to the development of swamps.
See: Permafrost, Taiga.

Turbidity

The term given by scientists to the increase in aerosols in the Earth's atmosphere.
See: Aerosols, Atmosphere, Global Warming.

Turbulence

In meteorology, this refers to the irregular flow (movement) of gases due to friction. Two types have been identified. Mechanical turbulence occurs when obstacles, either natural or man-made, on the surface of the Earth create eddies in the wind, thereby dissipating some of its kinetic energy and slowing it down. Thermal turbulence, such as convection, occurs because of the heating of the surface by solar radiation. Variable gusts of wind are an indication of turbulence.
See: Convection, Friction Layer.

Typhoon

The name given to a hurricane that develops over the waters of the western Pacific in tropical areas.
See: Hurricane.

Ubac *(Below)*

A French term relating to the slope of a mountain that is orientated north to south and, therefore, receives far less sunshine and light than a slope that runs from east to west, which is known as an adret. Both terms generally refer to slopes in the Alps.
See: Opaco.

Ultraviolet Radiation

This is the shortwave electromagnetic radiation emitted by the sun, most of which is prevented from reaching the surface of the Earth by the atmosphere. However, changes in the composition of the atmosphere may lead to potentially dangerous levels of ultraviolet radiation from reaching lower level, possibly leading to increases in the incidence of certain types of cancer.
See: Chlorofluorocarbons, Global Warming, Ozone Layer.

Umbra

The term used to describe either the central shadow of either the Earth of Moon during an eclipse, or the central dark area associated with a sunspot.
See: Penumbra, Solar Eclipse, Sunspot.

Unstable Air Layer

In an unstable air layer, a rising parcel of air stays warmer than the air through which it is ascending, and so continues to rise; while a descending parcel of air remains colder than the air through which it is falling and therefore continues to fall.
See: Atmospheric Stability, Stable Air Layer.

Upwelling

The process by which cold, deep oceanic waters rise to the surface.
See: El Nino.

Valley Breeze

A type of localized and temporary wind found in summer. Valley breezes reflect the differential heating that occurs between the slope of a mountain facing the sun and the floor of a deep valley. Freed from winter's snow cover, the valley wall heats up very quickly during the daytime, causing the air adjacent to the slope to warm and begin to rise. The air over the valley at the same altitude as the warming parcel of air is colder and therefore descends to replace the air that is following upslope from the valley floor to fill the area vacated by the rising warm air. Consequently, a circular pattern of air movement is initiated, which is strongest between the middle of the morning and sunset. During the night a reverse flow pattern is often established.
See: Mountain Breeze, Wind.

Vapor Pressure

As water vapor molecules mix with the other gases that make up the atmosphere, they add to the total pressure exerted on a body by the atmosphere. Vapor pressure is, therefore, the contribution that water vapor in the atmosphere makes to the atmospheric pressure and is directly proportional to the amount of water vapor in the atmosphere. As water vapor is approximately four percent of the lower level of the atmosphere, where the atmospheric pressure is accepted to be 1013.25 millibars, then it is possible to say that the vapor pressure at this lower level is about 40 millibars.
See: Absolute Humidity, Millibar, Specific Humidity.

Vardarac

A wind similar to the mistral experienced in the northwestern Mediterranean, the vardarac gains its name from the Vardar river in Macedonia. Cold winds of this type originate in the region's mountain ranges and blow forcefully down its valleys.
See: Mistral, Wind.

Veering

A wind is said to veer when it changes its path in a clockwise direction. The opposite, anticlockwise, change is called backing.
See: Backing, Wind

Vendavales

Most commonly associated with the eastern coastal region of Spain and Gibraltar, this is a strong and squally wind that is prevalent in winter and is accompanied by depressions bearing heavy rain. It is a southwesterly wind.
See: Depression, Squall, Wind.

Veranillo

The name given in many parts of tropical South and Central America to the short dry season which may break up the longer wet season.
See: Tropics, Verano.

Verano

The term used to describe the long dry season in parts of tropical South and Central America.
See: Tropics, Veranillo

Virazon

A form of sea breeze found on the west coasts of Chile and Peru in South America.
See: Sea Breeze, Terral, Wind.

Virga

Also known as fallstreaks, virga are associated with clouds and consist of trails of descending ice crystals or water droplets that evaporate before they have the opportunity to reach the surface. Virga often appear as hooked streaks of cloud, chiefly because as the water droplets or ice crystals evaporate as they fall, their speed decreases relative to those droplets or crystals just beginning their descent. Virga formations are found in conjunction with many types of cloud, including altocumulus, cirrus, and stratocumulus.
See: Atlocumulus, Cirrus, Cloud, Cloud Classification, Evaporation, Stratocumulus.

Visibility

A general term usually defined as the maximum distance at which an object can be seen clearly. Many types of weather feature can restrict visibility, including dust particles, fog, heat haze, pollution, and precipitation.
See: Fog, Precipitation.

Viscosity

Friction, the resistance that one body meets when it moves when in contact with another body, affects not only solids, but also fluids (liquids and gases). When friction is applied to a fluid, it is called viscosity. There are two types: molecular viscosity, which denotes the random movement of the molecules that comprise a liquid or gas; and eddy viscosity, which develops within fluids due to external influences. With regard to wind, eddy viscosity affects a wind when it comes into contact with objects such as walls, buildings, and vehicles. Eddies, located on the leeward (down-wind) side of such an obstacle, are indicative of turbulence and a slowing of the wind, usually leading to the depositing of any particles, such as dust or snow, blown along by the wind.

See: Friction Layer, Turbulence.

Volcano (Below right)

While there can be no doubt about the potentially destructive forces generated by a volcanic eruption, climatologists are unclear as to the influence that such episodes have on the climate and weather of the Earth. Interest has focused on two aspects: the degree to which eruptions influence changing patterns over time, and the precise nature of the possible changes generated by volcanic activity. Historical evidence suggests that volcanic eruptions lead to an overall cooling of the atmosphere. For example, the eruption of Krakatoa in the Indonesian archipelago during 1883 was followed by several years with cooler than average mean temperatures. Scientists initially concentrated on studying the huge volumes of dust and ash particles that are thrown out into the atmosphere during an eruption, suggesting that they increased the reflective powers of the planet and thereby reduced the amount of solar radiation that could reach the surface. However, it became clear that the majority of particles that are ejected in a spectacular eruption are generally too heavy to stay in the atmosphere for any length of time and were unlikely, therefore, to have any pronounced effect on climate in either the medium or long term. More recently attention has concentrated on the gases that pass into the atmosphere during an eruption. Once it has passed into the stratosphere, sulfur oxide gas combines with atmospheric moisture to produce sulfuric acid and sulfate particles, collectively known as sulfurous aerosols. These tend to remain aloft in the atmos-

phere for some considerable time due to their small size and the stratosphere's lack of precipitation which, at lower levels, tends to wash pollutants out of the atmosphere. These sulphurous aerosols in the upper atmosphere react with solar radiation, absorbing some of it and therefore reducing the amount that is transmitted to the surface. Consequently, while the stratosphere is warming, the troposphere is cooling. Over time, the layer of sulfurous aerosols in the stratosphere has become more pronounced, partly due to a succession of volcanic eruptions. In recent years, one of the most significant eruptions was that of Mount Pinatubo in the Philippines, which began on June 12, 1991. Vast quantities of sulfur dioxide was pumped into the atmosphere, and meteorologists believe that this produced a noticeable drop in the Earth's average temperature which lasted for the following two years. However, Mount St Helens in Alaska in 1980 produced no such climatic change, because the gases thrown out had only low concentrations of sulfur oxides. However, localized changes in temperature were noted by those recording the aftermath of the event.

See: Aerosols, Atmosphere, Climatic Change, Global Warming, Sulfur Dioxide

Wake

This refers to the zone of turbulent air that forms on the leeward side of obstacles such as buildings, windbreaks, or natural features.
See: Friction Layer, Leeward, Turbulence, Viscosity.

Wall Cloud

A rotating cloud feature associated with a meso-cyclone, a well-developed thunderstorm that may evolve into a tornado. Wall cloud is an approximately circular cloud that is found at the base of a thunderstorm in the area where the strongest updrafts of air are found. This updraft brings in humid but rain-cooled air into the lower regions of the mesocyclone, which expands and cools further as the air rises into the system. Condensation within the rising air begins to produce cloud at a level that is below that of the base of the cloud associated with the thunderstorm.
See: Condensation, Mesocyclone, Thunderstorm, Tornado.

Warm Air Advection

Advection refers to the movement of an air mass from one position to another, and warm air advection takes place when an air mass which has developed over a warm area is blown by the prevailing winds into a colder area.
See: Advection, Air Mass, Cold Air Advection.

Warm Front

This is the line on the Earth's surface where an advancing mass of warm air is forced to rise by the presence of a cooler, slower-moving, and denser air mass. Cloud develops as the warm air is forced aloft and precipitation is likely to take place ahead of the front. Once the front has passed, temperatures tend to rise, precipitation ends, and the wind veers. Warm fronts are most frequently found in high latitudes during winter and are linked to the development and decay of depressions.
See: Cloud, Cold Front, Depression, Occlusion, Warm Sector.

Warm Sector

This refers to the area of warm air found situated between a cold and warm front. Linked to a depression, the warm sector has low cloud and light rain, and persists until the advancing cold front catches up with the warm front to form an occlusion.
See: Cold Front, Depression, Warm Front, Occlusion.

Warm Wave

A sudden rise in temperature in temperate zones, usually in summer when warm air arrives from lower latitudes in advance of a depression moving eastward or to the west of an anticyclone.
See: Anticyclone, Depression, Temperate Zone.

Water

Water has three phases: solid, liquid, and vapor, and changes between the various states can either add latent heat to the environment or remove it. Ice melting or water evaporating gain latent heat from the environment, as does sublimation, the change of ice directly to a vapor. Vapor condensing to a liquid and water freezing to a solid release latent heat into the environment, as does deposition, the change of vapor to a solid without going through the intermediate phase of being a liquid. Water moves between the three phases due to heat or the lack of it
See: Cloud, Deposition, Evaporation, Hydrological Cycle, Precipitation, Sublimation, Water Vapor.

Water Cycle

The movement and exchange of water in solid, liquid, and gaseous form between the atmosphere, the land, vegetation, and bodies of water. Not all of the Earth's water takes part in the cycle at any one time. For example, underground stores of water such as aquifers and permanent ice sheets essentially remove water from the cycle; while the cooling or warming of the Earth can either remove or add water to the cycle by increasing or decreasing the volume held in glacier or ice masses.
See: Hydrological Cycle, Water, Water Vapor.

Waterspout *(see illustrations pp 183 and 184/5)*

Essentially, this is a tornado that develops at sea, most frequently in tropical or sub-tropical regions. A waterspout consists of an inverted cone of cloud descending toward the surface of the water from a cumulonimbus cloud formation. The descending cone of cloud encourages spray to rise from the surface water which forms into a rotating column. The upper regions of a waterspout tend to move faster than its lower levels, so a waterspout tends to become deformed, and eventually breaks up.
See: Cloud, Cloud Classification, Cumulonimbus, Tornado.

Water Vapor

A phase of water in which the water is a gas. Water can change to a gas in two ways: by sublimation during which it goes from a solid to a gas, without going through the intermediate phase of being a liquid, or by evaporation, during which water goes directly from a liquid to a gas. In both cases, the water gains latent heat from the environment within which the phase changes are taking place. As the concentration of water vapor in a parcel of air increases, so the density of the parcel decreases and the concentration of water vapor in a parcel of air is referred to as its humidity.
See: Absolute Humidity, Air Mass, Atmosphere, Evaporation, Relative Humidity, Specific Humidity, Sublimation, Water.

Wave Cloud

See: Cloud, Cloud Classification, Lenticular Cloud.

Weather

This is the condition of the Earth's atmosphere at a particular point in time or over a short period, and is identified from various meteorological recordings, including temperature, amount and type of

precipitation, amount of cloud cover and sunshine, and wind speed and direction. Weather exhibits change over short periods unlike climate which refers to changes occurring seasonally or over many thousands of years.

See: Climate, Climatology, Cloud, Meteorology, Precipitation, Seasons, Sunshine, Wind

Weather Chart

This is a regional, national, or international map which show various aspects of the weather at a particular point in time. The data is a compilation of weather data provided by weather stations or weather-recording devices spread throughout the area covered by the chart. These maps are also known as synoptic charts because they provide a synopsis of prevailing conditions at a particular period. Such maps usually carry information relating to systems that are responsible for the weather at a particular place. These include fronts and areas of high and low pressure. High pressure areas are identified by the letter H, while low pressure areas are identified by the letter L. The symbols used to located and identified a front depend on the nature of the front. Cold fronts are represented by a blue toothed line (the

position of the teeth indicates the direction of movement); warm fronts consist of a red line punctuated by semicircles, and a stationary front comprises a mixture of both. Charts can also contain indicators, each with their own symbol, of precipitation, pressure readings, wind speed and direction, and temperature values. Rain is generally symbolized by a dot, drizzle by a comma, and snow by a star, although these are not always employed by forecasters, particularly on television broadcasts.

See: Cold Front, Drizzle, Front, Rain, Snow, Warm Front, Weather.

Weather Ship

A manned vessel fitted with a wide array of instruments for recording various aspects of the weather at sea, such as pressure, temperature, wind speed and direction, and precipitation. They are usually anchored at a fixed position. Their data is of particular value to fishing fleets.

(Weather Buoy illustrated page 186)

Weather Vane *(Below left)*

Similar to a wind vane, this is a device for locating the direction of the prevailing wind.

See: Wind Sock, Wind Vane.

Wedge of High Pressure

A modified form of a ridge of high pressure in which isobars have a distinctive v-shaped appearance.

See: Ridge of High Pressure, Trough of Low Pressure.

Westerlies

Winds found in both the southern and northern hemisphere that exhibit variation in both their strength and direction. They are found along the poleward edges of the Horse Latitudes and are southwesterly in the northern hemisphere and northwesterly in the southern hemisphere. In both cases they blow away from the high pressure zones associated with tropical anticyclones located in belts at 30 degrees north and south of the equator. The westerlies are important components of the weather experienced in the midlatitude temperature zones.

See: Anticyclones, Horse Latitudes, Polar Front, Temperate Zone, Wind.

Wet Spell

In Great Britain, the term given to a period of 15 days or more in which each daily rainfall total has reached or exceeded one millimeter.
See: Dry Spell.

Whirlwind

A small-scale, localized weather feature consisting of a column of rising and rotating warm air carrying dust into the atmosphere.
See: Dust Devil, Duststorm, Tornado.

Whiteout

This is the name given to extreme weather conditions, usually found during violent snowstorms or blizzards, when both manmade structures and natural features of the landscape are either invisible or indistinguishable from each other.
See: Blizzard, Snow.

Williwaw

The name given to the strong and frequently occurring westerly squalls found in the Strait of Magellan, lying between South America and Tierra del Fuego , which connects the Atlantic and Pacific Oceans in the southern hemisphere. It is also used to describe a cold wind blowing downslope in Alaska.
See: Squall.

Willy-Nilly

An Australian term referring to the tropical cyclones active in the northwest of the country. These develop off the warm Timor Sea and then move southward before tracking back to the southeast and crossing the Australian coast. A typical willy-nilly is associated with heavy rain and tends to die away as it passes into the interior of the country.
See: Cyclone.

Wind

Wind is defined as the movement of air in relation to the rotation of the Earth and varies in its speed and direction, both horizontally (between two points) or vertically (with altitude). Wind direction is also identified with the direction from which a wind blows, so a north wind is a wind which is blowing from (not to) the north. Winds can be localized, regional, and planetary wide, and can

be shortlived or nearly continuous. There are several interlinked forces that have an influence on both the direction and speed of wind: air pressure gradients, centripedal force, coriolis force, friction, and gravity. Air masses have different pressure properties. Warm air has a higher pressure than cold air and air flows from areas of high pressure to zones of lower pressure. The greater the contrast in the two pressures, the greater the velocity of the wind. Centripedal force develops when a wind follows a curved path and influences changes in direction of a wind. Coriolis force influences the direction in which a wind moves, deflecting major winds to the left of their initial direction in the southern hemisphere and to the right in the northern hemisphere. Its impact increases as winds move farther away from the equator. Friction, a product of obstructions, both natural and manmade, on the Earth's surface, effects lower-atmosphere winds by slowing them down, while gravity impacts on the rate of downward motion of air. Wind direction and speed then is determined by the interaction between these various factors. Air also exhibits vertical changes in pressure, with pressure generally higher at lower altitudes. Consequently air flows from low altitudes to higher altitudes, but only if the gradient is sufficiently strong to overcome the descending influence of gravity. At higher altitudes, horizontal winds are influenced by horizontal pressure gradients and the coriolis force. These geostrophic winds follow a straight path when the pressure gradient and coriolis force achieve an equilibrium. Horizontal gradient winds are large-scale, higher altitude features, which follow a curved path and are associated with either anticyclones or cyclones.
See: Air Mass, Air Pressure, Anemometer, Anticyclone, Beaufort Scale, Chinook, Coriolis Force, Cyclone, Doldrums, Fohn, Friction Layer, Geostrophic Wind, Hadley Cell, Horse Latitudes, Intertropical Convergence Zone, Jet Stream, Land Breeze, Lake Breeze, Mountain Breeze, Polar Easterlies, Polar Front, Pressure Gradient, Rossby Waves, Trade Wind, Valley Breeze, Westerlies.

Windbreak

Any type of device for reducing the speed of the prevailing wind, which can consist of a row of trees or a fence, for example. Wind blowing

through a windbreak is forced into eddies on the leeward side of the barrier, thereby dissipating some of its energy and slowing down. Any matter, such as snow or dust, carried by a wind will be deposited on the leeward side of a windbreak.
See: Friction Layer, Leeward, Turbulence, Viscosity, Wind, Windward.

Wind Chill

This term refers to the cooling effect of low temperatures and wind on the human body. Generally, as wind speed increases, it becomes more effective in taking heat and water away from the surface of the body by, respectively, conduction and evaporative cooling.
See: Conduction, Evaporation.

Wind Rose

This is a diagram which shows the proportion of prevailing winds which blow from a particular direction over a set period of time. The simplest wind rose consists of a series of lines converging around a central circle; the position of each line indicates the direction of the wind, while their lengths show the rough proportion of the winds blowing from that direction. The longer the line, the greater the proportion of the prevailing wind from that particular direction.
See: Wind.

Wind Sock *(Right)*

A device often found at airports: usually an open ended, cone-shaped bag attached to a tall pole, which shows the direction of the prevailing wind. It also gives a generally indication of wind speed. As the speed rises, the sock fills with moving air and becomes increasing horizontal to the surface.
See: Wind Vane.

Wind Vane

This is a simple device used to indicate the direction of the wind. A typical vane comprises a moving horizontal arm with a direction pointer at one end and a fixed frame showing the four cardinal points of the compass: north, south, east, and west. As the wind changes direction, the horizontal pointer moves and, by comparing its position with the four cardinal points, the wind's direction can be identified.
See: Wind.

Winter *(Illustration page 190/1)*

One of the four generally recognized seasons of the year found in the Earth's temperate zones. In the northern hemisphere, winter falls between December and January, while in the southern hemisphere it falls between June and August.
See: Season, Temperate Zone.

Windward

The side of an object, such as a building or mountain, or the direction which faces the prevailing wind.
See: Leeward, Wind.